Do Elephants Jump?

Also by David Feldman

How Do Astronauts Scratch an Itch?

What Are Hyenas Laughing At, Anyway?

How Does Aspirin Find a Headache?

When Did Wild Poodles Roam the Earth?

Do Penguins Have Knees?

Why Do Dogs Have Wet Noses?

When Do Fish Sleep?

Who Put the Butter in Butterfly?

Why Do Clocks Run Clockwise?

How to Win at Just About Anything

Imponderables®

Do Elephants Jump?

An Imponderables® Book

David Feldman

Illustrated by Kassie Schwan

Collins
An Imprint of HarperCollins*Publishers*

HarperCollins books may be purchased for educational, business, or sales promotional use. For information, please write: Special Markets Department, HarperCollins Publishers, 10 East 53rd Street, New York, NY 10022.

First Collins edition published 2005.

Designed by Joseph Rutt

The Library of Congress has catalogued the hardcover edition as follows:
Feldman, David
Do elephants jump? / David Feldman; illustrated by Kassie Schwan.—
1st ed.
p. cm.
"An Imponderables® book."
ISBN 0-06-0539135 (hc)
1. Questions and answers. I. Schwan, Kassie. II. Title.
AG195.F439 2004
031—dc22 2004042890

ISBN-10: 0-06-053914-6 (pbk.)
ISBN-13: 978-0-06-053914-3 (pbk.)

05 06 07 08 09 ❖/RRD 10 9 8 7 6 5 4 3 2 1

For James Gleick

CONTENTS

PREFACE

Now we know how exterminators feel.

Almost twenty years ago, we took it as our mission to help eradicate annoying pests from our world. But despite our nine previous books, Imponderables still spring from dark corners like cockroaches in a crumb-filled kitchen.

Imponderables are the little mysteries of life that drive us nuts until we find the solution—mysteries that other reference books won't tackle. Just as exterminators' wallets are fattened by the indomitable spirit of vermin and the phobias of big humans about little creatures, so we are lucky to have found a career stamping out pests one at a time. We may snuff out one Imponderable, but for every mystery vanquished, it seems as if another appears. In 1986, we figure out why some pistachios are dyed red. In 2004, we discover why orange juice tastes so awful after you brush your teeth.

Since our first book, the biggest change at Imponderables Central is the increasing reliance on the Internet. We now receive many more Imponderables submissions by e-mail than snail mail, and of course the World Wide Web is available for research. An exterminator might utilize space-age chemicals but still has to get on his hands and knees to spray the crack under the floorboard. Likewise, we find that we can't rely on the expertise of random Web Sites. We use the Web, but mostly as a place to find the same caliber of experts that we've always relied upon to answer even the most elusive Imponderable.

To celebrate our tenth Imponderables® book, and because it's been several years since the last volume, we're devoting more space to answering your Imponderables (every single one of our Imponderables in this book came from a reader, and the first person to ask each of the published questions wins a free, autographed copy) and to your letters, even if most of them are taking us to task for perceived malpractice. And, responding to countless requests, we have included a master index to all ten Imponderables® books and *Who Put the Butter in Butterfly?*

Since the last book, there have been two developments we think you'd like to know about. One is that Malarky®, a game based on the Imponderables® books, is available across North America. And www.imponderables.com is now the cyberspace outpost of all things Imponderable. It includes news and information about what's happening at Imponderables Central, a blog written by Dave Feldman, and absolutely no banners or pop-up ads.

One thing hasn't changed since the first book, *Imponderables*. Your involvement is crucial to the fun. The last few pages of the book will let you know how to get in touch with us. But just remember: We vanquish Imponderability, not creepy six-legged things. Imponderables: Let's get ready to rumble!

Why Was He Called the Lone Ranger When Tonto Was Always Hanging Around?

The classic western features a lone hero entering a new town and facing a villain who threatens the peacefulness of a dusty burg. The Lone Ranger, on the other hand, came with a rather important backup, Tonto. Leaving aside questions of political correctness or racism, calling the masked man the *Lone* Ranger is a little like calling Simon and Garfunkel a Paul Simon solo act.

Before we get to the "Lone" part of the equation, our hero actually was a ranger, in fact, a Texas Ranger. *The Lone Ranger* started as a radio show, first broadcast out of Detroit in 1933, created by George Trendle, and written by Fran Striker. The first episode established that circa 1850, the Lone Ranger was one of six Texas Rangers who were trying to tame the vicious Cavendish Gang. Unfortunately, the bad guys ambushed the Rangers, and all of the Lone Ranger's comrades were killed. The Lone Ranger himself was left for dead. Among the vanquished was the Lone Ranger's older brother, Dan.

So for a few moments, long enough to give him his name, the

Lone Ranger really was by himself. He was the *lone* surviving Ranger, even if he happened to be unconscious at the time. Tonto stumbled upon the fallen hero and, while nursing him back to health, noticed that the Ranger was wearing a necklace that Tonto had given him as a child. Many moons before, the Lone Ranger (who in subsequent retellings of the story we learn was named John Reid) saved Tonto's life! Tonto had bestowed the necklace on his blood brother as a gift.

When Reid regained his bearings, the two vowed to wreak revenge upon the Cavendish Gang and to continue "making the West a decent place to live." Reid and Tonto dug six graves at the ambush site to make everyone believe that Reid had perished with the others, and to hide his identity, the Lone Ranger donned a black mask, made from the vest his brother was wearing at the massacre. Tonto was the only human privy to the Lone Ranger's secret.

Not that the Lone Ranger didn't solicit help from others. It isn't easy being a Ranger, let alone a lone one, without a horse. As was his wont, Reid stumbled onto good luck. He and Tonto saved a brave stallion from being gored by a buffalo, and nursed him back to health (the first episode of *The Lone Ranger* featured almost as much medical aid as fighting). Although they released the horse when it regained its health, the stallion followed them and, of course, that horse was Silver, soon to be another faithful companion to L.R.

And would a lonely lone Ranger really have his own, personal munitions supplier? John Reid did. The Lone Ranger and Tonto met a man who the Cavendish Gang tried to frame for the Texas Ranger murders. Sure of his innocence, the Lone Ranger put him in the silver mine that he and his slain brother owned, and turned it into a "silver bullet" factory.

Eventually, during the run of the radio show, which lasted from 1933 to 1954, the duo vanquished the Cavendish Gang, but the Lone Ranger and Tonto knew when they found a good gig. They decided to keep the Lone Ranger's true identity secret, to keep those silver bullets flowing, and best of all, to bounce into television in 1949 for a nine-year run on ABC and decades more in syndication.

The Lone Ranger was also featured in movie serials, feature movies, and comic books, and the hero's origins mutated slightly or weren't mentioned at all. But the radio show actually reran the premiere episode periodically, so listeners in the 1930s probably weren't as baffled about why a law enforcer with a faithful companion, a full-time munitions supplier, and a horse was called "Lone."

Submitted by James Telfer IV of New York, New York.

When Does a Bill Become a Beak on a Bird?

There's no ceremony when proud parents beam as their warbler's bill graduates into beakdom. In fact, there's no difference at all between a bill and a beak. They are one and the same.

The relative size of birds' bills varies enormously from species to species, and bills are much more instrumental to a bird than our proboscis, to which it is sometimes compared by the avian ignorant. The beak of a bird is a bony organ that surrounds the mouth and is essential to birds' ability to eat food in the wild—depending upon the bird, the bill can serve as a chef's knife, fork, food processor, or serving plate. For example, the hard, conical bill of a sparrow is designed to crush seeds, while hawks' bills are hooked to facilitate tearing the flesh of their prey, and hummingbirds have long, thin bills to probe delicate flowers for nectar.

Sometimes scientists can be killjoys. For some reason, ornithologists prefer to use the term *bill.* Why anyone would prefer the bland *bill* to the cool *beak,* we can't figure out, but a look at the scientific literature will confirm what Allison Wells, communications director of the prestigious Cornell Laboratory of Ornithology told us:

We use only *bill* around the Lab. Though *bill* and *beak* refer to exactly the same thing, *bill* is the more proper term, and it's the one we use. However, you will see *beak* used occasionally in some literature on birds.

Indeed, veterinarians treat a serious disease called psittacine beak and feather disease. In the less scientific bird press, you'll see references to "beaks," especially when discussing birds with large bills, such as flamingos, pelicans, or parrots.

Just like we tend to apply "beak" to humans with large schnozzes, so do birds with large mandibles receive the more colorful appellation. But to say the least, confusion reigns. We remember from this old limerick by Dixon Lanier Merritt as starting with these words:

A wonderful bird is the pelican
His beak can hold more than his belican

Look up "The Pelican" on the Internet and you'll see the limerick with *bill* at least as often as *beak*, while the rest of the limerick is identical.

Submitted by Mark Kramer and Kevin McNulty of San Diego, California.

What Does "Legitimate" Theater Mean? Where Can You Find "Illegitimate" Theater?

Call us grumpy, but we think laying out a hundred bucks to listen to a caterwauling tenor screech while chandeliers tumble, or watching a radical reinterpretation of *Romeo and Juliet* as a metaphor for the Israeli-Palestinian conflict is plenty illegitimate. But we are etymologically incorrect; the use of the word *legit* dates back to the end of the nineteenth century, when it was used as a noun to describe stage actors who performed in dramatic plays. It soon became a term to describe just about any serious dramatic enterprise involving live actors.

And to this day, "legitimate" is used to describe actors who toil in vehicles that are considered superior in status to whatever alternatives are seen as less prestigious. As Bill Benedict of the Theatre Historical Society of America points out, one of the definitions of *legit* in *The Language of American Popular Entertainment* is:

> Short for *legitimate*. Used to distinguish the professional New York commercial stage from traveling and nonprofes-

sional shows. The inference is that *legit* means stage plays are serious art versus popular fare.

Back in the late nineteenth century when the notion of "legit" was conceived, live public performances were more popular than they are today, when television, movies, the Internet, DVDs, and spectator sports provide so much competition for the stage. Even several decades into the twentieth century, other types of amusements, such as minstrel shows, vaudeville, burlesque (with and without strippers), magic shows, and musical revues often gathered bigger crowds than legitimate theater.

"Illegitimate" actors had a shady reputation, as most were itinerant barnstormers who swept in and out of small or medium-sized towns as third-rate carnivals do today. Their entertainments tended to be crude, with plenty of pantomime, caricature, low comedy, and vulgarity, so as to play to audiences of different educational levels, ethnicities, and even languages.

Cleverly, promoters of "legitimate" theater appealed to elite audiences, who could afford the relatively expensive tickets and understand the erudite language. Theater critics emerged well into the nineteenth century in the United States, trailing behind the British, who already featured theater reviewers in newspapers. The more affluent the base of the newspapers, the more critics would tend to separate the "mere" entertainments from the aesthetic peaks of serious theater.

These cultural cross currents are still in play today. Theater critics in New York bemoan the "dumbing down" of Broadway shows, Disney converting animated movies into theater pieces, and savvy producers casting "big name" television or movie stars in plays for their marquee value. And the stars are willing to take a drastic reduction pay in order to have the status of legitimate theater bestowed upon them; they appear on talk shows and proclaim, "My roots are in theater." We've yet to see a leading man coo to an interviewer: "My roots are in sitcoms."

Not everyone takes these distinctions between "legit" and "illegit" so seriously. When Blue Man Group, with its roots in avant-garde theater, brought its troupe to the Luxor Hotel in Las Vegas, Chris Wink, cofounder of the Blue Man Group, proclaimed: "Now that Vegas has expanded its cultural palette and embraced Broadway-style legitimate theater, it feels like a good time to introduce some illegitimate theater."

Submitted by Carol Dias of Lemoore, California.

Why Do Pianos Have 88 Keys?

Our pianos have a peculiar configuration, with 52 white keys and 36 black keys, ranging from A, 3 ½ octaves below middle C, to C, four octaves above middle C. Why not 64 keys? Why not 128?

Before there were pianos, there were pipe organs. In medieval times, some pipe organs included only a few keys, which were so hard to depress that players had to don leather gloves to do the job. According to piano historian and registered piano technician Stephen H. Brady, medieval stringed instruments originally included only the white keys of the modern keyboard, with the raised black keys added gradually: "The first fully chromatic keyboards [including all the white and black keys] are believed to have appeared in the fourteenth century."

Clavichords and harpsichords were the vogue in the fifteenth and sixteenth centuries, but they kept changing in size and configuration—none had more than four octaves' range. Octave inflation continued along, as the ever more popular harpsichord went up as high as a five-octave range in the eighteenth century.

In 1709, a Florentine harpsichord builder named Bartolomeo Cristofori invented the pianoforte, an instrument that trumped the harpsichord by its ability to play soft (*piano*) or loud (*forte*) depending upon the force applied on the keys by the player. Brady notes that the first pianos looked very much like the harpsichord but

> were fitted with an ingenious escapement mechanism which allowed the tones to be produced by tiny hammers hitting the strings [the mechanism attached the hammers to the keys], rather than by quills plucking the strings as was the case in the harpsichord.

Others soon created pianos, but there was little uniformity in the number of keys or even in the size of the piano itself.

Michael Moore, of Steinway & Sons, theorizes that it was a combination of artistic expression and capitalism that gave rise to the 88-key piano. Great composers such as Mozart were demanding instruments capable of expressing the range of the music they were creating. Other composers piggybacked on the expanded range provided by the bigger, "modern" pianos. Piano makers knew they would have a competitive advantage if they could manufacture bigger and better instruments for ambitious composers, and great changes were in store between 1790 and 1890, as Stephen Brady explains:

> By the end of the eighteenth century, toward the end of Mozart's career and near the beginning of Beethoven's, piano keyboards had reached six full octaves, and a keyboard compass of six and a half octaves was not uncommon in early nineteenth-century grands. For much of the middle to late nineteenth century, seven full octaves (from lowest A to highest A) was the norm. A few builders in the mid-nineteenth century experimented with the seven-and-a-quarter-octave keyboard, which is in common use today, but it did not become the de facto standard until about the 1890s.

Steinway's grand pianos had 85 or fewer keys until the mid-1880s, but Steinway then took the plunge to the 88 we see today, and other manufacturers rushed to meet the specifications of their rival. But why stop at 88? Why not a nice, round 100? Michael Moore explains:

> Expansion into still greater numbers of keys was restrained by practical considerations. There is a limit to the number of tones that a string can be made to reproduce, especially on the bass end, where low notes can rattle, as well as a limit to the tones that the ear can hear, especially on the treble end. There is a type of piano, a Boesendorfer Concert Grand, which has 94 different keys, [and a full eight-octave range, with all six of the extra keys added to the bass end], but by and large our 88 keys represent the extent to which pianos can be made to faithfully reproduce tones that our ears can hear.

Even if more keys would gain the slightest advantage in tones, there is also the consideration of size and weight. The Boesendorf is almost ten feet in length, exceeded only by the ten-feet, two-inch Fazioli Concert Grand. Only a handful of compositions ever ask to use these extra keys, not enough reason to motivate Boesendorfer to add the keys in the first place. According to Brady, "The Boesendorfer company says the extra strings are really there to add sympathetic resonance and richness to the regular notes of the piano's range."

Submitted by Guy Washburn of La Jolla, California.

Why Do Rice Cakes Hold Together?

Our correspondent, duly reading the ingredient list on his package of rice cakes, notes that only rice and salt are listed. He rightfully wonders how rice cake makers manage to keep together what would seem to be fragile rice. Is there a secret binding ingredient in the mix?

We're sure the rice cake producers would say that the secret ingredient is love, but emotion has nothing to do with it. We contacted several rice cake producers and received the same explanation from all of them (a rarity in the Imponderables business) about how rice cakes are formed.

First, uncooked rice is soaked in water and then mixed with a little salt (and in some cases, with a bit of oil). This soaking is important, because the moisture from the rice is going to help puff it up when it is heated in the grain-popping machine, as Quaker Foods and Beverages explains:

> A rice cake is formed when heat and pressure are added to the grain, causing it to expand abruptly. A portion of grain is set onto a round, metal pan—like a mini–baking pan. As a hot cylinder presses down onto the pan, sizzling pressure is released. The heat is so intense that after only a few seconds, the grain makes a loud popping noise as it bursts. This process causes the grains to "pop" and interweave. There are no oils, additives, or binding ingredients used during this process.

If the rice cake is flavored, the seasonings are applied after the popping process, and doesn't affect the sticking together of the rice itself.

Rice cakes date back to 3000 B.C. in Southeast Asia, and home cooks have never been privy to the specialized equipment that modern commercial rice cake makers enjoy. Home cooks in Asia make rice cakes by soaking glutinous rice overnight, steaming the rice until it is soft, grinding the heck out of it with a mortar and pestle, and then pounding the mashed rice with a mallet. Then they knead the rice like bread dough and cook it, resulting in a rice cake (or rice ball) with a smoother consistency than that of Western cakes.

Whether using the traditional methods or specialized metal molds designed only for rice cake production, bakers seem to have no trouble getting rice cakes to hold together—now if only they could manage to produce some taste!

Submitted by Dane Bowerman of Muir, Pennsylvania.

Why Doesn't the Water in Fire Hydrants Freeze During the Winter?

There may be no such thing as a dumb question, but there are certainly ones that are based on false assumptions. The water doesn't freeze in hydrants for the same reason that the water in empty ice cube trays doesn't form cubes. You can't freeze what's not there!

Bob Ward, former president of the SPFAAMFAA (Society for the Preservation and Appreciation of Antique Motor Fire Apparatus in America), told *Imponderables* that there is a nut at the top of a hydrant that controls the flow of water to the hydrant. When a firefighter wants to open the hydrant, a wrench is used to open the nut. Attached to the shut-off nut is a long stem that goes to the valve at the bottom of the hydrant, underground, that controls the flow of water. When the water flow is closed, the standing water above the valve is drained automatically. As Ward succinctly puts it, "There is no water to freeze."

When a firefighter is finished with the hydrant, he or she closes the nut and the water drains below the shut-off valve automatically. The shut-off valve is located well below the frost line, so fire hydrants rarely encounter any freezing problems, even in lovely climes like Chicago's or Oslo's.

Submitted by Todd Sanders of Holmdel, New Jersey

Why Do So Many Bars Feature Televisions with the Sound Turned Off?

We spare no financial expense, no mental duress, in order to plumb the depths of Imponderability. To research this question, we tore ourselves away from the plush confines of Imponderables Central to visit many taverns. Risking inebriation and worse, we confirmed that the "Yes, we have a TV on; no, we don't have the sound on" phenomenon is alive and well in North America. What's the deal?

Somewhat to our surprise, we found bartenders uniformly negative about the boob tube and its role in their establishments. Why does management bother installing televisions? The thinking seems to run on the order of:

> Where there are bars, there are men.
> Where there are men, there is an interest in sports.
> Sports is televised.
> Sports on television equals male butts on our stools.

If we don't have televisions at our bar, men will go to the sports bar down the street instead.

But the bartenders we spoke to analyzed this Imponderable more deeply. Televisions are important because they provide patrons with what Dan Sullivan, a Kiwi now living and bartending in Greece, calls "something to do with their eyes." Single patrons are often uncomfortable and tense when alone. They may be lonely, or worried about looking like losers, or anxious about meeting potential mates. The television "makes it easier for them to be by themselves at the bar," concludes Roger Herr, owner of South's Bar in downtown Manhattan.

Some bars and nightclubs also use televisions to run closed-circuit programming, anything from old Tom and Jerry cartoons to 1960s-style light shows to help set the appropriate mood for their establishments. One bartender compared this use of the television to installing fish tanks, a form of visual Muzak.

Every bar employee we talked to indicated that as soon as the audio on a television goes on, some patrons are turned off. As Deven Black, former manager of the North Star Pub in New York City, put it,

> No matter how quietly the sound is on, it will offend someone, and you can never have it loud enough so everyone who wants to can hear it.

Even manly men might not want to accompany their scotch-and-sodas with the mellifluous tones of NASCAR engines backfiring. And bartenders reported that most sports fans are perfectly content with the audio of their sports programs on mute, happily shedding commercials and colorless color commentators.

All nightclubs and most bars feature music, whether a humble jukebox, live bands, or expensive sound systems. If the TV is going to interfere with the music, why pump dollars into the jukebox? If customers are going to listen to Marv Albert instead of Bruce Springsteen, what owner is going to be happy about installing a $20,000 sound system?

But most of the bar industry folks we consulted make a more spiritual point. As bartender and beer columnist Christopher Halleron put it,

> People go to bars for conversation and socializing. When you turn up the boob tube, that element is taken away as people become fixated on whatever it spews and stop talking to each other. The same phenomenon occurs in the living room of the average American family.

Exactly! If we wanted to sit sullenly and watch blinking images while avoiding human contact, we'd stay at home with our families.

Liquor flows more freely when patrons feel festive, and music and dancing set the mood more easily than *Wheel of Fortune* or *Everybody Loves Raymond.* A blaring television sucks the energy out of a room.

Some bars have used modern technology to solve the television-audio problem in a Solomon-like way: they turn on the closed-captioning option on their TVs. CC might not be the solution if patrons are trying to hear the New York Philharmonic on PBS, but then, they never are.

Submitted by Fred Beeman of Las Vegas, Nevada.

Why Do So Many Taverns Put Mirrors in Back of the Liquor Bottles Behind the Bar?

No doubt, many tavern owners install mirrors in the back of their bars for the same reason most businesses do anything—because their competitors are doing the same thing. We were surprised that some bar owners couldn't explain why they have mirrors behind their bars, but most of the same folks who weighed in on the last Imponderable had plenty of opinions about this one, too.

Like a television, a mirror provides patrons something to look at when they might feel lonely, tense, or bored. And there can be more practical advantages, as Deven Black notes: "It allows patrons to check each other out discreetly."

Sometimes, the view might not be so pleasant ("Uh-oh, here comes my girlfriend! And I told her I'd be home at eight."), but more than a century ago, some bars ensured that the view would be more pleasing to their clientele, as Gary Regan reveals,

> In the late 1800s, a "naughty" painting of [William Bouguereau's] *Nymphs and Satyr* hung in New York's Hoffman House bar. It was situated so that customers could stare at it through the mirror, therefore not blatantly looking at a naughty painting. Presume, therefore, that mirrors were and are used [by customers] to observe the scene without being obvious.

Bartenders are not impervious to using the mirror for less than professional purposes, as an honest but lascivious bartender and beer columnist Christopher Halleron explains:

> Mirrors provide an excellent, indirect way to check out the cleavage on the girl who just ordered a Cosmo.

Although it may be a surprise to some libidinous patrons, there are things to look at in a bar other than beef- or cheesecake, and mirrors are an inexpensive accent piece in a tavern's interior decoration. For one, mirrors bounce light around the room, and can also be attractive themselves. Many beer and liquor companies provide free mirrors with their name and logos emblazoned on them. Some bars prefer to use the mirrors to advertise themselves. Tom Hailand, design engineer at Cabinet Tree Design, believes that mirrors add "glitz," and even point-of-purchase advertising on the mirrors can yield practical benefits to the bar:

> And where else would you put your sandblasted logo just in case the customers are so hammered they need the name of the bar in front of them to call a cab?

Heiland also mentions an advantage to the placement of the mirror that was echoed by many experts—the mirror makes the liquor display look fuller. Mirrors have long been used by decorators to make rooms look bigger and displays more enticing. As one bartender told us,

> The idea is to make the liquor more appealing by spicing up the presentation of the bottles as well as making it appear that there are more bottles than there really are. It's impressive when you walk into a bar and see a huge shelf full of liquor bottles behind the bar—the mirror provides the same effect with fewer bottles. The same trick is used in the catering business for veggie trays and other food presentations."

But the predominant reason for mirrors in bars, and probably the precipitating factor in the tradition's beginning, was for security. Think back to the Old West, and it's easy to see why a saloon owner would want advance warning before a gunslinger, with pistol packed, entered the establishment. There were times when it was unavoidable for the bartender to turn his or her back to patrons, as bartender "Baudtender" wrote:

> Many of the old bars had the bartenders' "make station" (where they prepared mixed drinks) right below the liquor storage shelves. With the mirrors, they could keep an eye on the customers while their back was turned—before Dram Shop laws [which made bartenders legally liable for harm in-

flicted by intoxicated or underage patrons], it was a common thing to give a customer an entire bottle of liquor and charge by measurement or eyeball estimation for what was consumed. Contrary to popular opinion, bar owners weren't the only rogues to water down the liquor, if you see what I mean.

Most cash registers at bars are located in back of the bar, so bartenders must turn their back on patrons on a regular basis.

Today, there are other dangers, large and small, that prompt a bartender to use the mirror. Without the mirror, the bartender might ignore the quiet patron who would otherwise go unnoticed, or miss the guy trying to steal the bartender's tip while the bartender's back was turned, or fail to assist the woman about to be harassed. What's especially appealing about using the mirror for security purposes, according to Roger Herr, owner of South's Bar in Manhattan, is that the bartender can scan the area without being obvious.

And just in case the bar brawl should erupt, bartender Dan Morrison sings the praise of mirrors for providing the élan that their manufacturers wouldn't trumpet but movie stunt coordinators are well aware of:

> Mirrors smash really good when you throw a chair at them.

Submitted by Charlie Chiarolanza of Lafayette Hill, Pennsylvania.

Why Does the Pope Change His Name upon Assuming His Office?

The tradition of name changes for Church officials dates back to the beginnings of the Christian movement. In the Book of Matthew, Jesus appoints his disciple Peter to be the first head of the Church. Yet Peter's birth name was Simon.

According to Lorraine D'Antonio, the retired business manager of the Religious Research Association,

> Their names were changed to signify the change in roles, attitudes, and way of life. When the Pope assumes his role as head of the Church, it is assumed that he will change his life as his name is changed when he dedicates his total being to the service of the Church. Quite often, when a person dedicates his or her life to the service of God (during ordination, confirmation, etc.), the person assumes a new name, as he or she assumes a new role and commitment.

The first pope who we know changed his name was John II, in A.D. 533. Presumably, his given name, Mercurio, a variation of the pagan god Mercury, was deemed unsuitable for the head of the Church. So Mercurio paid tribute to John I by adopting his name.

No other Pope changed his name until Octavian chose John XII for his papal name in 955. A few decades later, Peter Canepanova adopted John XIV. Brian Butler, president of the U.S. Catholic Historical Society, told *Imponderables* that John XIV was the first of several popes with the given name of Peter to start a tradition—no pontiff has ever taken the name of the Church's first pope.

Butler notes that the last pope to keep his baptismal name was Marcellus II, born Marcello Cervini, who served in the year 1555.

Submitted by David Schachow of Scarborough, Ontario.

Do Identical Twins Have Identical Fingerprints? Identical DNA?

Let's put it this way: If identical twins had identical fingerprints, do you really think David Kelley wouldn't have fashioned a murder plot about it on *The Practice*? Or Dick Wolf on one of the seventy-four *Law and Order* spinoffs?

Scientists corroborate our TV-based evidence. Identical twins result when a fertilized egg splits and the mother carries two separate embryos to term. The key to the creation of identical twins is that the split occurs *after* fertilization. The twins come from the same sperm and egg—and thus have the same DNA, the identical genetic makeup.

Because their DNA is an exact match, identical twins will always be the same sex, have the same eye color, and share the same blood type. "Fraternal twins," on the other hand, are born when two separate eggs are fertilized by *two different* sperms. Their DNA will be similar, but no more of a match than any other pair of siblings from the same parents. Not only do fraternal twins not necessarily look that much alike (think, for example, of the Bee Gees brothers, Maurice and Robin Gibb), but can be of the opposite sex and may possess different blood types.

Even though identical twins share the same DNA, however, they aren't carbon copies. Parents and close friends can usually tell one identical twin from the other without much difficulty. Their personalities may differ radically. And their fingerprints differ. If genetics doesn't account for these differences, what does? Why aren't the fingerprints of identical twins, er, identical?

The environment of the developing embryo in the womb has a hand in determining a fingerprint. That's why geneticists make the distinction between "genotypes" (the set of genes that a person inherits, the DNA) and "phenotypes" (the characteristics that make up a person after the DNA is exposed to the environment). Identical twins will always have the same genotype, but their phenotypes will differ because their experience in the womb will diverge.

Edward Richards, the director of the program in law, science, and public health at Louisiana State University, writes:

> In the case of fingerprints, the genes determine the general characteristics of the patterns that are used for fingerprint

classification. As the skin on the fingertip differentiates, it expresses these general characteristics. However, as a surface tissue, it is also in contact with other parts of the fetus and the uterus, and their position in relation to uterus and the fetal body changes as the fetus moves on its own and in response to positional changes of the mother. Thus the microenvironment of the growing cells on the fingertip is in flux, and is always slightly different from hand to hand and finger to finger. It is this microenvironment that determines the fine detail of the fingerprint structure. While the differences in the microenvironment between fingers are small and subtle, their effect is amplified by the differentiating cells and produces the macroscopic differences that enable the fingerprints of twins to be differentiated.

Influences as disparate as the nutrition of the mother, position in the womb, and individual blood pressure can all contribute to different fingerprints of identical twins.

Richards notes that the physical differences between twins widen as they age: "In middle and old age [identical twins] will look more like non-identical twins."

About 2 percent of all births are twins, and only one third of those are identical twins. But twins' fingerprints are like snowflakes—they may look alike at first blush, but get them under a microscope, and the differences emerge.

Submitted by Mary Quint, via the Internet. Thanks also to Rachel P. Wincel, via the Internet; and Stephanie Pencek, of Reston, Virginia.

Why Does Monopoly Have Such Unusual Playing Tokens?

What do a thimble, a sack of money, a dog, a battleship, and a top hat have in common? Not much, other than that they are among the eleven playing tokens you receive in a standard Monopoly set. And don't forget the wheelbarrow, which you'll need to carry all that cash you are going to appropriate from your hapless opponents.

The history of Monopoly is fraught with contention and controversy, for it seems that its "inventor," Charles Darrow, at the very least borrowed liberally from two existing games when he first marketed Monopoly in the early 1930s. After Darrow self-published the game to great success, Parker Brothers bought the rights to Monopoly in 1934.

On one thing all Monopoly historians can agree. When Parker Brothers introduced the game in 1935, Monopoly included no tokens, and the rules instructed players to use such items as buttons or pennies as markers. Soon thereafter, in the 1935–1936 sets, Parker Brothers included wooden tokens shaped like chess pawns: boring.

The first significant development in customizing the playing pieces came in 1937, when Parker Brothers introduced these die-cast metal tokens: a car, purse, flatiron, lantern, thimble, shoe, top hat, and rocking horse. Later in the same year, a battleship and cannon were added, to raise the number of tokens to ten.

All was quiet on the token front until 1942, when metal shortages during World War II resulted in a comeback of wooden tokens. But the same mix of tokens remained until the early 1950s, when the lantern, purse, and rocking horse were kicked out in favor of the dog, the horse and rider, and the wheelbarrow. Parker Brothers conducted a poll to determine what Monopoly aficionados would prefer for the eleventh token, and true to the spirit of the game, the winner was a sack of money.

Parker Brothers wasn't able to tell us why, within a couple of years, Monopoly went from having no tokens, to boring wooden ones to idiosyncratic metal figures. Ken Koury, a lawyer in Los Angeles who has been a Monopoly champion and coach of the official United States team in worldwide competition, replied to our query:

> Monopoly's game pieces are certainly unique and a charming part of the play. I have heard a story that the original pieces were actually struck from the models used for Cracker Jack prizes. Any chance this is correct?

We wouldn't stake a wheelbarrow of cash on it, but we think the theory is a good one. We contacted author and game expert John Chaneski, who used to work at Game Show, a terrific game and toy emporium in Greenwich Village, who heard a similar story from the owner of the shop:

> When Monopoly was first created in the early 1930s, there were no pieces like we know them, so they went to Cracker Jack, which at that time was offering tiny metal tchotchkes, like cars. They used the same molds to make the Monopoly pieces. Game Show sells some antique Cracker Jack prizes

and, sure enough, the toy car is exactly the same as the Monopoly car. In fact, there's also a candlestick, which seems to be the model for the one in Clue.

John even has a theory for why the particular tokens were chosen:

> I think they chose Cracker Jack prizes that symbolize wealth and poverty. The car, top hat, and dog [especially a little terrier like Asta, then famous from the "Thin Man" series] were possessions of the wealthy. The thimble, wheelbarrow, old shoe, and iron were possessions or tools of the poor.

Submitted by Kate McNieve of Phoenix, Arizona. Thanks also to Mindy Sue Berks of Huntington Valley, Pennsylvania; Flynn Rowan of Eugene, Oregon; and Sue Rosner of Bronx, New York.

What Are Those Black Specks on Corn Chips, Tortillas, and Tortilla Chips? And Why Are the Specks on Nacho Cheese Tortilla Chips Less Dark?

When you buy fresh sweet corn, throw away the husk and silk, and gaze upon the kernels of yellow or white corn, not a black speck is in sight. Boil the corn in a pot or grill them on the barbecue, and no specks appear. So why, when we gaze upon a Frito or grab a mound of Doritos, can we see "freckles" on every single chip? We've noticed that the specks seem dark on most corn chips and yellow corn tortilla chips, and lighter but still easily visible on cheese-flavored chips and plain white corn tortilla chips (the difference between yellow and white corn is due to a few pigments in the outer layers of the corn kernel).

We called the folks at Frito-Lay, which makes both of these brands, and they swore up and down that they did nothing to add specks to the chips. Frito-Lay uses whole corn kernels and grinds and cooks them in-house to make Fritos and Doritos. The customer service representative, who preferred to remain anonymous, assured us that the specks are not pepper. They are not burn marks or signs of an overcooked chip (if so, then every chip they sold would be burnt). "So what are they, then?" we demanded. "Errrr . . ." she responded.

Actually, this Imponderable stumped a bunch of experts, including professors and researchers who specialize in corn. We called Irwin Steinberg, our old pal and executive director of the Tortilla Industry Association, who despite his many years in the business, professed that he has never been asked about the black specks that are also evident in corn (but not flour!) tortillas or tortilla chips. But he knew where to send us: to two professors "who'll know the answer to your questions."

Our heroes are Lloyd Rooney, chairman of the Food Science faculty and Ralph Waniska, professor of soil and crop sciences at Texas A&M. We grilled them separately and both told us the same story, so either they are foisting an elaborate hoax on us, or this Imponderable has been solved!

On the tip of each corn kernel is a hilum, collectively known as the "black layer," where it is attached to the cob. While corn is growing, nutrients are being transferred from the rest of the cob to the kernels through the hilum. The hilum serves a function for the corn not unlike a belly button attached to a human umbilical cord serves for a human fetus—it's a nutritional supply line to the rest of the kernel.

But the hilum is not always black. As the corn matures, the hilum turns from unpigmented to light green to light brown to brown and eventually to black. When the corn is fully mature, the hilum is black and nutrients are no longer being sent to the kernel.

If mature corn forms a black layer, and every kernel has a black tip, then why don't we see it in sweet corn? The answer is that sweet corn for the home table is picked more than a month before the corn used by the chip companies, before the hilum discolors.

When corn is starting to grow, most of the kernel is composed of moisture. As they mature, the kernels gain starch and lose some of their "milk." Sweet corn is picked when the corn is at maximum sweetness, and there is still plenty of moisture within the kernel. Fresh sweet corn commands a premium price. But industrial food processors would rather wait longer to pick corn for tortilla chips or feed grain for cattle, because as corn continues to lose moisture and gain more starch, it provides more usable product.

Some chip companies, rather than using the whole corn and creating the masa themselves (which involves boiling corn in a lime solution, rinsing, draining, grinding, and then drying the corn and forming it into a doughlike consistency), buy prepared dry masa flour and the black specks aren't as prevalent. If the flour was made with the outer layer of the kernel removed, there may be no specks at all. Too bad, because as Rooney says, "the hilum is probably a good source of fiber and perhaps antioxidants and is not detrimental in any manner."

Indeed, some consumers, hip to the hilum, equate a lack of specks with inauthentic chips. So some crafty snack food companies actually buy the hila (the plural of hilum) separately and add them to the dough to pass off their chips as though they were "made from scratch." We cry foul!

Why do white tortilla chips have fainter specks than yellow ones? Because white corns have lighter hila.

And why do cheese-flavored yellow tortilla chips, such as Nacho Cheese Doritos, sport lighter specks? Because the flavorings are partially obscuring the specks. The cheese is like a cosmetic palliative—a speck concealer.

Submitted by Melissa Taylor of Holland, Michigan. Thanks also to Dot Finch of Soddy-Daisy, Indiana.

Why Do We Say "P.U." When Something Smells Awful?

We were wondering if the etymology dates back to one Pépé le Pew's debut on the silver screen. Perhaps the spray of a real skunk motivated the first exclamation of "P.U.," but a trip to any dictionary will confirm that this expression of displeasure long predated animated cartoons. Somewhat to our surprise, *P.U.* is not an abbreviation and not an acronym.

In Latin, the word *puteo* means "to stink, be redolent, or smell bad." The Indo-European word *pu* refers to rot or decay, and many other languages contain words referring to bad smells that start with the letters *pu*. The English interjection *phew* refers to "a vocal gesture expressing impatience, disgust, discomfort, or weariness" according to the *Oxford English Dictionary,* and variants abound (*pew, pho, pheut, phoo, phugh, peugh,* and *fogh,* dating back as far as the early seventeenth century).

What all these variations have in common, according to Jesse Sheidlower, the North American editor of the *Oxford English Dictionary,* whose "Word of the Day" column is a highlight of the *OED*'s

Web site, is that all represent "something like the whistling sound you get by blowing a puff of air out of closed lips." Sheidlower's theory, as good as any we can come up with, is that "P.U." is an emphatic form of saying "pew."

> Other words can be given emphasis in a similar way: *kee-rist!* for "Christ," or *bee-yoo-ti-ful* for "beautiful," are two examples. The difference with *P.U.* is that it has the advantage of sounding like two named English letters, which is why it's spelled out that way, instead of something like *pee-yew.*

No one knows for sure when the disyllabic pronunciation of *P.U.* has been around. Sheidlower says the spelling "certainly existed by the 1950s, if not earlier," and most agree that the United States can claim ownership of the short, if not sweet, expression.

Submitted by Susan Sales of Pittsburgh, Pennsylvania. Thanks also to Noelle Yamada of Honolulu, Hawaii; Camille A. Buckley of Elma, New York; Leslie Wingard of Jacksonville, Florida; and Bruce Greeley of Fall City, Washington.

Do Elephants Jump?

We talked to a bunch of elephant experts and none of them has ever seen an elephant jump. Most think it is physiologically impossible for a mature elephant to jump, although baby elephants have been known to do so, if provoked. Not only do mature elephants weigh too much to support landing on all fours, but their legs are designed for strength rather than leaping ability. Mark Grunwald, who has worked with elephants for more than a decade at the Philadelphia Zoo, notes that elephants' bone structure makes it difficult for them to bend their legs sufficiently to derive enough force to propel the big lugs up.

Yet there are a few sightings of elephants jumping in the wild. Veterinarian Judy Provo found two books in her college library that illustrate the discrepancy. S. K. Ettingham's *Elephant* lays out the conventional thinking: ". . . because of its great weight, an elephant cannot jump or even run in the accepted sense since it must keep one foot on the ground at all times." But an account in J. H. Williams's *Elephant Bill* describes a cow elephant jumping a deep ravine "like a chaser over a brook."

Animals that are fast runners or possess great leaping ability have usually evolved these skills as a way of evading attackers. Elephants don't have any natural predators, according to the San Diego Wild Animal Park's manager of animals, Alan Rooscroft: "Only men kill elephants. The only other thing that could kill an elephant is a fourteen-ton tiger."

Most of the experts agree with zoologist Richard Landesman of the University of Vermont, that there is little reason for an elephant to jump in its natural habitat. Indeed, Mike Zulak, an elephant curator at the San Diego Zoo, observes that pachyderms are rather awkward walkers, and can lose their balance easily, so they tend to be conservative in their movements.

But that doesn't mean that elephants are pushovers. Why bother jumping when you can walk through or around just about everything in your natural habitat? In India, trenching has been the traditional way of trying to control movements of elephants. Veterinarian Myron Hinrichs, of Petaluma, California, notes that the traditional trench has to be at least two meters deep, two meters across at the top, and one and one-half meters across at the bottom to serve as a barrier for elephants:

> That tells us that they can't or won't try to jump a distance of 6.5 feet. But these trenches have a high failure rate, for elephants can fill them in, especially in the rainy season, and then

walk across the trough they have made. And larger bull elephants can go down through and up even a trench that size.

Why leap when you can trudge?

Submitted by Jena Mori, of Los Angeles, California.

What Is the Purpose of the Recessed Notch on the Bottom of Most Round Shampoo Bottles? Why Are There Two Notches on Some?

When we received this Imponderable, we picked up a couple of shampoo bottles and noticed that the deep ends of the notches are aligned with the seams on the bottles. We therefore assumed that the notch was a way of reinforcing the seam, or an innocent by-product of affixing the seam.

Wrong! The little notch actually has a purpose. We feel a little less dumb about not knowing, since Dr. John F. Corbett, ex–vice president of scientific and technical affairs at Clairol, and now consultant to the company, didn't know the answer himself. He thought the notches might be there to aid in releasing the bottle from the mold. But luckily, Corbett is curious, and shared the real deal with us:

> The recessed notch or notches are there to allow orientation of the round bottle when it is being labeled by silk screening. By orienting the bottle, the print can be applied in such a way as to avoid the seam being within the label area.
>
> Non-round bottles (oblong or oval cross sections) can be aligned by making use of the narrower front-to-back dimension as they pass along the labeling line. A bottle with a perfectly square cross section would need to have orientation notches.

There is no significance as to whether one or two notches are used. Incidentally, the seam results from the join between the two halves of the mold in which the bottle is formed.

Submitted by George C. Lady IV of Hayward, California.

Why Do Police Officers Hold Flashlights with an Overhand Grip?

Reader Raphael Klayman writes:

I've noticed that police (real ones as well as those who play them on TV) hold their flashlights the way one might hold a knife to stab someone in the chest. We civilians tend to hold our flashlights from underneath, in a kind of semi-bowling or fishing rod grip. With both methods, you can shine the flashlight from floor to ceiling, but the police style feels a lot more awkward. Yet they must have their reasons. What are they?

Although few of the police officers we contacted have ever received formal training in flashlight "gripology," the overhand style was the favored position for two main reasons. The most often cited rationale was alluded to in our reader's letter: The overhand grip allows the officer to use the flashlight as a weapon. One Tennessee officer wrote us:

The way we hold a flashlight gives you a tactical advantage against the person that you are encountering in case of a use-of-force situation. Your arm is already in a raised or almost a defensive position against an attack, not to mention you have something in your hand that can strike a pressure point.

With just the quick flick of a wrist, an officer might be able to stop an unruly perpetrator; if the flashlight were held straight in front, the hand would have to be drawn back first.

The second reason for the "police grip" is that perhaps its most common use is to survey the occupants and contents of motor vehicles. In conventional cars, the officer is above the level of the driver and the car, and it is simply more comfortable to beam the flashlight downward in the overhand grip. As one cop put it:

> When I am walking up to the car, I don't think I ever make eye contact for more than a few seconds at a time. I'm looking through the car, at hands, containers, etc. You cannot see those things well by holding the flashlight at waist level.

What about when an officer is searching a darkened house? We finally found one officer who preferred the underhand grip for this purpose, with his arm out in front of him, "like I would when I'm holding a gun." But most officers sided with this viewpoint:

> I never extend my arms out unless I intend on shooting someone. If I'm moving through a house with my arms extended and a burglar is around the corner, and he sees my arms and flashlight, there is a pretty good darn chance he might grab me. If your arms are already extended out, they are bound to get tired. There is no strength in those arms if they are extended; the closer the hands and arms are to the body, the stronger they are going to be.

Once firearms are added to the mix, flashlight position can become a matter of life and death. An Arizona police officer wrote:

> We are trained to hold our flashlights in a specific way when firing our weapons. I was taught many years ago to always place your flashlight in your "weak hand" so that you would be

> able to pull your weapon quickly if needed. The reason that we hold them in a "stabbing" type manner is that they are easier to bring up to the firing position if needed. By holding the flashlight in a "stabbing" position by the switch, the officer is able to bring the flashlight up toward the target, and it allows the officer to use the back of his or her strong hand against the back of his or her weak hand in order to support the proper alignment in case of need to fire the weapon.

Many police departments use lightweight Maglite flashlights, and many companies that manufacture firearms for law enforcement agencies provide mounts for putting the Maglites directly on the weapon, as one officer explained:

> I have a tactical flashlight attached to the Mag tube of my Mossberg that comes in handy for investigating suspicious noises around the barn and workshop at night. The light also makes a good aiming guide, because the shot is centered on the circle of light.

And with the light source attached to the firearm, all those nasty decisions about how to grip the flashlight are moot.

Submitted by Raphael Klayman of Brooklyn, New York.

Why Are Newspapers So Effective in Cleaning Windows?

If you read most newspapers for more than a few minutes, your hands feel dirty. So why do windows look squeaky clean when you rub the same newspaper against the glass? We asked experts in window cleaning, newspaper printing, and inks this very same question.

One potential allure would have to be price. Newsprint, the kind of paper used for newspaper printing, is probably the cheapest paper manufactured—and discarded newspaper the cheapest of all. Still, we couldn't find a single professional cleaner who uses newspapers on the job, although several could see why newspapers could be effective in a pinch. Jim Grady, of Tri-State Window Cleaning in Wappinger Falls, New York, writes:

> I don't use newspaper to clean windows. I use squeegees. When I was a child, my father did tell me that newspapers were an effective way to clean windows, probably because it was free, or cheaper than paper towels. But as the good book

says, when I became a man I put away my childish ways and I haven't put a piece of newspaper on glass in twenty-five years.

The first quality that is prized in a drying agent is absorbency, and in this regard, newspaper hits the jackpot. Theodore Lustig, at the West Virginia University School of Journalism, told us:

> Newsprint is extremely porous [larger spaces between fibers] and is uncoated [no waxes or fillers]. Therefore, it pretty much acts like a sponge, making it useful in whatever tasks require sopping up moisture. This may account for its ability to clean windows.

Another proponent of the absorbency theory is Jim Patton, process manager at Smurfit Newsprint, in Pomona, California, who adds that newsprint's lack of water repellency is another reason for its success in cleaning windows:

> Newsprint is pretty darn absorbent. Most papers have a sizing applied to them to give them water repellency. Newspapers have a minimum of repellency. At a different newspaper manufacturer I worked for, they used newsprint in the men's room for drying your hands.

Of course, the lack of sizing in newsprint isn't to aid in window cleaning—it's to please newspaper publishers. Bob Cate, director of manufacturing services at newsprint producer Bowater, Inc., agrees that the lack of sizing in newsprint is key to its efficacy in cleaning windows, and explains how:

> You want any ink to soak into the paper, as opposed to standing out on the surface, but you don't want it to soak in so fast that it comes out on the other side. This is why newspaper

is absorbent—so that the ink can soak in and not sit on top of the paper when it's applied.

But it gets better! Newspapers not only are cheap and absorbent, but they give the cleaned, dry windows a shine, according to Brent Weingard, of Expert Window Cleaner (and evidently a part-time comedian):

> Being a New York window cleaner, I have found the *Daily News* to be the popular favorite, especially on Midtown storefronts. In the East Village, the *Village Voice*. My upscale residential clients, however, are more comfortable if I arrive with the *New York Times* or the *Wall Street Journal* in hand.
>
> Seriously, to my knowledge, newspaper is generally never used by professional window cleaners. It's really more of a "home remedy" cleaning tool that actually works surprisingly well. It doesn't have lint, is very absorbent, and gives the glass a shiny finish.
>
> My theory of what causes that shiny finish is evidenced by looking at your hands after handling newspapers: it's ink! I believe it is a film of ink that is left on the glass surface that gives windows this colorful reflective finish.

So what are the downsides of professionals using newspapers to clean windows? They are few, but they are prohibitive. The first, enunciated by Richard Fabry, publisher of *American Window Cleaner* magazine, is the killer:

> Sure, newspaper gets the glass clean, but it is s-l-o-w. Get a professional squeegee and it'll go much faster. Window cleaning should be quick 'n' easy.

Another problem: What do you do with the newspapers when you are finished with them? Gary Mauer, of Window Cleaning Network in Oconomowoc, Wisconsin, a town with more *o*'s in its name than window cleaners, writes:

> Because we clean so many windows, it's simply not practical to use and dispose of bundles of newspaper. For that reason alone, you'll be hard pressed to find a professional window cleaner who uses newspaper.

Next time you clean your own windows, you might want to give newspapers a try instead of paper towels. The pros tended to look askance at paper towels—they dry the glass surface well enough, but leave lint—cloth diapers or tight-knit towels or rags are a better bet. With newer inks, often vegetable- rather than petroleum-based, newspapers tend to streak less than in the past (and for that matter, stain your hands much less when you read them). In years past, amateurs who used newspaper to clean windows tended to dirty white frames around the windows, streaking them with ink.

Submitted by B. Craig Sanders of Piscataway, New Jersey. Thanks also to John A. Beton of Chicago, Illinois; Jason Hsu, via the Internet; and Michael Gerstmann, via the Internet.

Why Do We Fly Flags at Half-Mast to Honor the Dead?

Although we now mourn dignitaries by flying the flag at half-mast on dry land, all of our sources agree that the custom has British naval origins. Prior to half-masting, ships would sometimes fly a black flag to honor a death. No doubt the old custom had at least two disadvantages: it necessitated bringing two sets of flags, one of which was unlikely to be used; and it failed in its primary purpose of signaling to others—a ship in the distance was much more likely to recognize a flag sailing at half-mast than to discern the color of a flag.

Although there are reports of flags at half-mast as early as the fourteenth century, the first recorded instance of half-masting occurred in 1612. An Eskimo killed Englishman William Hall, who was searching for the Northwest Passage (in what we now call Canada), and the Royal Navy lowered its flags at sea to honor him.

By the mid-seventeenth century, the Royal Navy had adopted the custom more formally, and its fleet flew its flags at half-mast annually to honor the death of King Charles I. Later, a commanding officer's death would lead to flying a ship's flag at half-mast, but eventually any crew member's passing would prompt a tribute.

Why honor the dead by lowering the flag? In his 1938 book *Sea Flags*, British Commander Hillary P. Mead speculates that the origins of half-masting date back to a deliberate attempt to make the boat as slovenly as possible, the opposite of shipshape:

> Untidiness and slovenliness of appearance were supposed to be the signs of grief, and this writer refers to biblical customs (amongst which may be mentioned sackcloth and ashes), and to the fact that in the Merchant Service ropes are left trailing and yards are scandalised in furtherance of this principle. This idea of slovenliness, at any rate in modern times, has no counterpart on land.

Another theory speculates that in the seventeenth century, regimental flags were placed on the ground when the Royal Family or foreign heads of states were saluted; merchant ships dipped their ensigns to warships. As Mead puts it:

> It naturally follows that national flags should be lowered as a salute to the departed, and remain lowered for a length of time proportionate to the importance of the deceased person.

As more and more countries mimicked the British custom at sea, half-masting became common on land throughout the world. But not everywhere. The Flag Research Center, based in Winchester, Massachusetts, mentions one exception in its *Flag Bulletin*:

> Because the religious inscription appearing on the national flag of Saudi Arabia is considered holy, the etiquette of that country forbids the flag being flown at half-staff, vertically, or upside down. Other forms of mourning have been used for national flags, including the addition of a black stripe at the fly end, a border of black on the three free edges, and the placement of black streamers on the pole above the flag.

Submitted by Laura Stone Bell of University Heights, Ohio. Thanks also to Howard Kim, via the Internet; and Jeanne Salt of Tualatin, Oregon.

Where Is the Donkey in Donkey Kong?

For those of you who didn't have anything better to do than obsess about Nintendo in the early 1980s, Donkey Kong is a game created by Shigeru Miyamoto, the most famous video game creator on the planet. Donkey Kong featured a diminutive hero, Jumpman (whose name was later changed to Mario), who had a much larger pet, a gorilla. The gorilla did not exactly bond with his "master," and conveyed his wrath by kidnapping Jumpman's girlfriend, Pauline, climbing a building, and hurling barrels and other missiles as our hero attempted to rescue his sweetheart. If the little man managed to reclaim her temporarily, the gorilla snatched Pauline away again. As the game progressed, each level made it harder for Jumpman to succeed. But regardless of what level the player progressed to, nary a donkey was seen.

So why the donkey in the title? Although some fans insist that the "donkey" was a misheard or mistranslated attempt at "Monkey Kong,"

Miyamoto has always insisted otherwise. On his tribute site to Miyamoto (http://www.miyamotoshrine.com), Carl Johnson includes an interview with Miyamoto at the Electronic Entertainment Exposition, where the game's creator addresses this Imponderable:

> Back when we made Donkey Kong, Mario was just called Jumpman and he was a carpenter. That's because the game was set on a construction site, so that made sense. When we went on to make the game Mario Brothers, we wanted to use pipes, maybe a sewer in the game, so he became a plumber.
>
> For Donkey Kong, I wanted something to do with "Kong," which kind of gives the idea of apes in Japanese, and I came up with Donkey Kong because I heard that "donkey" meant "stupid," so I went with Donkey Kong. Unfortunately, when I said that name to Nintendo of America, nobody liked it and said that it didn't mean "Stupid Ape," and they all laughed at me. But we went ahead with that name anyway.

In some other interviews, Miyamoto indicates that "donkey" was chosen for its usual connotation in English—stubbornness. In his book on Nintendo, *Game Over: Press Start to Continue,* David Sheff writes:

> When the game was complete, Miyamoto had to name it. He consulted the company's export manager, and together they mulled over some possibilities. They decided that *kong* would be understood to suggest a gorilla. And since this fierce but cute kong was donkey-stubborn and wily (*donkey,* according to their Japanese-English dictionary, was the translation of the Japanese word for "stupid" or "goofy"), they combined the words and named the game Donkey Kong.

At least one party wasn't happy with Nintendo's name—Universal Studios, which owned the copyright for *King Kong*. Universal sued for copyright infringement, claiming that the video game mimicked the basic plot of the movie (man climbs building to save his girlfriend from the clutches of a giant ape). Universal lost on the most obvious of grounds—the judge ruled that the movie studio did not own the rights to *King Kong*. Nintendo won the suit without, unfortunately, having to justify the nonexistence of a donkey in Donkey Kong.

Submitted by Darrell Hewitt of Salt Lake City, Utah.

Why Don't the Silver Fillings in Our Mouth Rust?

The fillings don't rust because there is no iron or steel in the amalgam, or "silver" fillings. Without iron, there is no rust.

But we understand the tenor of the question. Combine metal with constant exposure to air and liquid and you'd think your fillings would be devastated by corrosion. Here's why it isn't. There *is* silver in silver fillings, but it isn't the dominant component. Most amalgam fillings consist of approximately 50 percent mercury, with the rest silver, tin, copper, and zinc. Depending upon the alloy, the silver content can range between 2 and 35 percent, usually on the higher end of the scale.

Why is there more mercury than silver in fillings? Mercury has the ability to alloy with other metals. If you combined, say, silver, copper, and tin, you'd end up with a powder without any tensile strength. Mercury helps combine the other liquids to form a solid mass that is strong and yet can be compressed into a cavity and seal it effectively. Although the silver amalgam filling is inexpensive to manufacture and easy to install, the metal *does* corrode, which is one of the reasons why fillings sometimes have to be replaced. But the corrosion has a positive side, too, as Brooklyn dentist Philip Klein explains:

> When the restoration is inserted, a corrosive layer begins to form at the metal-tooth interface. This layer mechanically seals the restoration and prevents leakage that would ultimately lead to recurrent cavities and failure.

The corrosion actually prevents bacteria and other chemicals from entering the cavity-laden tooth.

All well and good, you might be saying, but isn't mercury a dangerous toxin? Yep, it sure is. Many countries in Europe, for example, have outlawed the use of amalgam fillings, yet the American Dental Association, the World Health Organization, and most mainstream health organizations maintain that the minimal amount of leakage of mercury from fillings is within acceptable guidelines for risk. Alternative and holistic dentists argue that mercury poisoning from amalgam fillings can cause everything from kidney damage to brain damage. Suddenly, a little corrosion in your mouth doesn't seem like a big deal.

Submitted by Claire Badger of Augusta, Georgia.

Why Do Beavers Build Dams?

Watch any nature documentary about beavers, and you'll see the giant rodents working furiously to construct their dams. Of course, if they weren't working furiously, they wouldn't deserve their hard-earned sobriquet. No one wants to be as busy as a sloth.

Beavers are industrious creatures, and we don't want to belittle their achievements, but we also don't want to fall into the trap of assuming that beavers cogitate deeply about how to solve their problems and build dams as a result. On the contrary, Dr. Peter Busher, professor of biology at the Center for Ecology and Conservation Biology at Boston University, told us:

> Most (if not all) beaver scientists would say that construction behavior is instinctual and not learned behavior. Thus construction behavior appears to be hard-wired into the beaver genome.

By damming small rivers and streams, beavers create ponds, still and deep bodies of water. Beavers create the pond by amassing dams with bases of mud and stones, and piling up branches and sticks. Beavers reinforce the dam by using mud, stones, and vegetation from the water as "plaster."

Scott Jackson, a wildlife biologist at the Department of Natural Resources Conservation at the University of Massachusetts, Amherst, told *Imponderables* that while some dams are less than 100 feet long, others have been recorded at over 1,500 feet in length. According to Jackson, beavers are constantly on the lookout for leaks, and will fix any defect by plugging leaks with mud and sticks.

Why do these rodents, who are not fish, after all, bother constructing ponds? Ponds help provide beavers with three of the necessities of life:

1. *Food.* Beavers are PETA-friendly strict vegetarians. Not only do they not eat other mammals, but they disdain fish and insects, as well. They do eat leaves, barks, and twigs of trees, and make meals from the vegetation that grows around the ponds they construct. By creating a pond, beavers increase the supply of aquatic vegetation, such as water lilies, that they prefer. As Dr. Busher notes,

> The higher water table favors plants that are water tolerant (many of which are eaten by beavers) and drowns species that

are not water tolerant. Beavers "feel" safer in water and the dams create ponds that allow them to have better access to food without leaving the water.

Why might beavers feel safer in the water? Because . . .

2. *Protection.* Most of beavers' predators, such as wolves, coyotes, and occasionally bears, are more comfortable on land than water (young beavers, called "kits," are also vulnerable to owls and hawks). Beavers are adept in the water: With their webbed hind feet and large paddle tails, beavers are built for swimming. Beavers are awkward on land, and most of their forays onto *terra firma* are to chew down trees and bushes both to eat and to use as construction materials. They are unlikely to wander far afield from the shoreline, where they are prone to be attacked by land-based predators.

3. *Shelter.* Obviously, beavers can't sleep under the water, so it is necessary for them to build a place to sleep safely atop the water. A family of beavers will frequently build a "lodge," a tepee-like structure usually made from tree branches, twigs, and aquatic vegetation sealed together with mud. Usually, lodges feature a bed of grass, leaves, and wood chips on which the beavers sleep and their young are raised—it's not Martha Stewart, but it's home.

The upper part of the lodge is built above the water line, and there is an opening at the top, so that breathing air and ventilation are available. According to Jackson, these lodges may be fifteen to forty feet across at the base and protrude three to six feet above the water. To stabilize the lodge, the beavers anchor the sides deep into the water and attach them to a solid structure such as an island in the water, a large underground branch, or sometimes even the side of the river. Beavers build tunnels from the air chamber of the lodge down to the water. These tunnels provide the only entrances to the lodge, as the thick walls of sticks and mud provide protection from both predators and the weather. By damming the river so that the pond is sufficiently large, the beavers create a sort of moat around the visible, above-water part of the lodge.

In the winter, if the pond freezes, the mud on the lodge freezes as well, providing more resistance to predators and bad weather. Even if a wolf were to walk on the frozen pond, it would struggle to penetrate the lodge, and beavers usually have plenty of time to swim through one of the tunnels to escape.

Even after a lodge is built, beavers must often perform repair work on the dams, as the water level of the pond is crucial to the safety of the lodge. If the water level gets too high, the lodge can submerge; if it gets too low, the lodge would become exposed at the bottom and predators could infiltrate the lodge more easily.

Whole families live in these lodges. Beavers are monogamous (they're too busy being busy as beavers to be promiscuous, we surmise), usually "marrying" for life barring the death of a mate. Females give birth inside of a lodge, usually producing four kits, but occasionally as many as nine. One family unit, called a "colony," usually consists of the two parents, that year's kits, plus the young from the previous year. As beavers are far from petite (North American adult beavers range in weight from thirty-five to eighty pounds), there's plenty of tail staying in one lodge at any given time.

Most animals adapt to their surroundings, but except for humans, no animal alters its habitat to suit its needs more than the beaver. The beaver may be able to convert part of a flowing stream into a still pond and base for its home, but it can't keep the same pond forever. Eventually, the beavers plunder the vegetation on the shoreline—what they don't eat, they use for building materials. If dams break, ponds can drain off. Where beavers live, silt tends to form, rendering the ponds too shallow. Any of these problems can cause beavers to abandon their lodges and seek other opportunities.

But all is not lost, as Scott Jackson reassures:

> In the nutrient-rich silt, herbaceous plants flourish, forming beaver meadows. Over time, shrubs and trees eventually come to dominate these areas, setting the stage for the beavers' return.

Submitted by Nathan Trask of Herrin, Illinois.

How Do They Keep Staples in Their Packages Clumped Together?

Ah, the irony. Although their product is used as a fastener, staple manufacturers must turn to a competitor to fasten the staples together. Those clumps of staples, more properly called "strips," are kept together by glue.

Staples are made out of wire. A machine called a "wire winder" wraps the wire around a spool. According to Lori Andrade, of staple manufacturer Stanley Bostitch, the cement is applied while the wire is wrapped around the spool. Then the flat wire is rolled off the spool, is cut into strips, and is pressed into that lovable staple shape.

Submitted by Karen Gonzales of Glendale, Arizona. Thanks also to Carlos F. Lima of Middleton, Wisconsin.

Why Did Pilgrims' Hats Have Buckles?

Would you be heartbroken if we told you that just about everything they taught you in elementary school about Pilgrims was wrong? We at Imponderables Central remember being forced to draw pictures of Pilgrims during elementary school, presumably to obscure the food stains on our families' refrigerators during the Thanksgiving season. We remember not liking to draw Pilgrims, because they wore boring black and white clothing, and the men wore those long black steeple hats sporting a gold or silver buckle.

So it is with more than a little feeling of righteous vengeance that we report that we were sold a bill of goods. Pilgrims might have worn hats, and those hats might have even been tall. But they were rarely black and never had a buckle on them.

How were generations brainwashed into thinking that Pilgrims wore buckled hats? For many Americans, there is confusion between the Pilgrims and Puritans. The two groups weren't totally unrelated: Both were early settlers in America in the early seventeenth century, and both groups fled England to escape what they considered to be an authoritarian and tyrannical Anglican Church, the state-sponsored religion of their government.

But in spirit, the two groups were far apart. The Pilgrims were separatists, who wanted to practice a simple religion without the rituals and symbolism that they felt had spoiled the "Protestant" church. Pilgrims first tried emigrating to Holland, but the poor economic conditions there, along with some religious intolerance, led one contingent to come to America. Approximately sixty of the one hundred passengers aboard the *Mayflower* were separatists (i.e., Pilgrims), and they settled in or near Plymouth, Massachusetts, in 1620.

Puritans, on the other hand, did not want to sever their relationship from the Anglican fold completely, but sought to "purify" the church. Puritans wanted to eliminate many of the reforms of the

Protestant movement, and return the church to more traditional practices. Several hundred Puritans moved to America in 1629, and settled the Massachusetts Bay Colony in what is now Cape Cod.

Both Puritans and Pilgrims have reputations as authoritarian, humorless, and conformist in their beliefs, but this stereotype characterizes the Puritans (who later in the century went on to conduct the Salem witch trials) more than the Pilgrims, who were much more democratic and inclusive in style. For example, the Pilgrims did, indeed, befriend local Native Americans, although it is unclear whether this pact was motivated by feelings of brotherhood or an arrangement for mutual self-defense.

Both Pilgrims and Puritans would probably be appalled that they are lumped together in Americans' consciousness today. Puritans would probably consider Pilgrims to be hopeless idealists, and too tolerant of dissent; Pilgrims would probably have deemed Puritans intolerant of others, and too timid to sever their links to the Anglican Church.

The two groups' different attitudes toward religion and democracy were reflected in their apparel choices. It was the Puritans who dressed the way Pilgrims are often depicted—with dark, somber clothing. Pilgrims, on the other hand, dressed much like their counterparts in England at the time. They did not consider it a sin to wear stylish or colorful clothing—indeed, several of the men who made the original trip on the *Mayflower* were in the clothing or textile trade. Many dyes were available to the Pilgrims, and they favored bright clothing—wills, provisions lists, written inventories, contemporaneous histories, and even sparse physical evidence all indicate that male *Mayflower* passengers wore green, red, yellow, violet, and blue garments along with the admittedly more common white, gray, brown, and black ones. The Pilgrim men wore many different types of hats, including soft caps made of wool or cloth, straw hats, and felt hats with wide brims. Wealthier Pilgrims might have worn more elaborate silk hats with decorative cords or tassels—but nary a buckle in sight.

We contacted Caleb Johnson, a Mayflower descendant who has

written the 1,173-page book *The Complete Works of the Mayflower Pilgrims* and hosts a Web site devoted to all things Pilgrim at www.mayflowerhistory.com. Johnson confirmed what we had read in other histories:

> The Pilgrims did not have buckles on their clothing, shoes, or hats. Buckles did not come into fashion until the late 1600s—more appropriate for the Salem witchcraft trials time period than the Pilgrims' time period.

So if Pilgrims didn't wear buckles, why have we always seen depictions of Pilgrims wearing what turns out to be nonexistent doodads on hats that they never actually wore? Johnson implicates writers:

> I am not sure I can pinpoint a specific reason as to why the popular image developed. I would suspect that authors and poets such as Nathaniel Hawthorne, Henry Wadsworth Longfellow, even Arthur Miller [in *The Crucible*], might have contributed to the "popular culture" image of a generic New England Puritan, which then got backward-applied to the early seventeenth-century Separatists—many not consciously realizing that 70 years separated the arrival of the Pilgrims and the more "traditional" Puritan we see portrayed at, say, the Salem witchcraft trials.
>
> It really wasn't until the mid-twentieth century that any serious scholarship into the archaeology, contemporary artwork, contemporary accounts, and analysis of historical records (such as probate estate inventories) of the Pilgrims enlightened us as to what they truly were wearing.

We also heard from Carolyn Freeman Travers, research manager and historian for the Plimoth Plantation, a "living-history museum" in Plymouth, Massachusetts ("Plimoth" was the preferred spelling of William Bradford, the first governor of the colony). The Plimoth

Plantation boasts a replica of the *Mayflower*, a re-creation of a Pilgrim village, arts and crafts, and no buckles. She dates the buckle-obsession to the early twentieth century, and thinks artists were the key perpetrators:

> The popular image of the Pilgrim developed in America about 1900 to 1920 into the man with the bowl-shaped haircut; tall, dark hat with the prominent square buckle; and large square buckles on his belt and shoes as well. The square buckles on the belt and shoes actually appear very frequently, and seem to mean quaint and "old-timey"—popular depictions of eighteenth-century people have them. Mother Goose, Halloween witches, and leprechauns generally do. The last often have the tall-crowned, narrow-brimmed hat with the buckle as well—it's the green color of the clothing that sets them apart.
>
> Why turn-of-the-century artists chose the buckle as a hat ornament to mean the Pilgrims/Puritans, I don't know. Earlier historical paintings of the mid-nineteenth century often had hats with a strap and buckle for Puritan men of the English Civil War. The famous 1878 painting by William Yeames, *And When Did You Last See Your Father?*, has a hat of this style, known as a sugar-loaf from the shape, with a strap and a buckle for the Puritan interrogator. There is also a similar hat in *The Burial of Charles I* (1857) by Charles W. Cope. My guess is that American painters looked to these paintings for inspiration, and went on from there.

Buckles did adorn hats in the late seventeenth century, though. We corresponded with the manager and curator of Plymouth's Pilgrim Hall Museum, Peggy Baker. Her museum features a cool Pilgrim-era felt hat processed from beaver furs that can be seen at www.pilgrimhall.org/beav_hat.htm. Baker concurred with our experts that buckles weren't around in the early seventeenth century, but came into vogue later:

> Buckles on hats were a genuine style, however—just not for Pilgrims. It was a short-lived style in the later seventeenth century, a fad, if you will. Why would anyone put a buckle on his hat? Who can really understand the vagaries of fashion? Imagine trying to explain logically and rationally to an audience 300 years from now the costumes worn, for instance, by Britney Spears and her imitators?
>
> Who could explain those costumes right now?

Baker agrees that artists are probably to "blame" for the buckle misconceptions:

> What happened, however, was that Victorian-era artists, illustrating the Pilgrims, were less interested in historical accuracy than in conveying the impression of "ye olde-timey." They used the buckled hat to convey that impression and the image became "stuck" in the popular imagination.

Submitted by Michael Goodnight of Neenah, Wisconsin.

Who Are All Those People on the Sidelines During American Football Games?

The action may be on the football field, but the traffic congestion is usually on the sidelines. In NFL games, but especially in big-college football schools, the area around the benches is teeming with as many people as Grand Central Station at rush hour. Who *are* all these guys? As Bob Carroll, executive director of the Pro Football Researchers Association puts it, the sidelines are full of

> . . . players, coaches, assistant coaches, equipment managers, towel boys, mascots, cheerleaders, officials holding the sticks, TV folks, photographers, police, alumni, anyone

donating big bucks to the school, and a partridge in a pear tree.

Restrictions on issuing credentials for access to the sidelines are surprisingly loose, especially in the pros. Faleem Choudhry, a researcher at the Pro Football Hall of Fame, told *Imponderables* that there isn't a hard and fast rule limiting the number of sidelines personnel, or even visitors: "Anybody the team deems necessary can be there." One team might want the electrician who supervises the lighting of the stadium to stay near the bench; another team might banish him to the stands.

The problem of overpopulated sidelines is greater in the college ranks, and the Big 10, known for its impassioned football competition, is among the most restrictive conferences in regulating credentials. The Big 10 allows a maximum of forty credentials for the bench area of each team, including all of the absolutely essential non-playing personnel, such as coaches, trainers, and physicians. According to Cassie Arner, associate sports information director of the University of Illinois, the bench area has a dotted line 50 yards long around it, usually starting at one 25-yard line and running to the other 25-yard line. The bench area zone does not extend all the way back to the stands, so cheerleaders and other credentialed personnel (in some cases, marching bands, press, and security) can stay behind the bench zone.

Here's how Arner estimates the University of Illinois allocates its credentials:

Ten to fifteen coaches
Approximately ten team managers (whose jobs range from handling balls to charting statistics for the team)
Five full-time equipment managers, who are responsible for mending damaged paraphernalia
Ten to fifteen trainers, of whom perhaps five are full-time doctors
The rest are student assistants there to get water, help with taping of bandages, and other relatively unskilled tasks

But other folks somehow manage to creep down to the bench area as well. In this category, Arner includes the team chaplain, security, and occasionally someone from the event management or operations department of the school. But the University of Illinois does not issue credentials for alumni. Occasionally, a big donor or a dignitary from another team might be brought down to the "forty zone" during time-outs or at the quarter breaks. An occasional "honorary coach" is given credentials—usually a professor from the university who has helped with recruiting.

Tom Schott, sports information director at Purdue University, concurs with his Illinois counterpart, although it sounds like Purdue is a little looser in issuing credentials. As he says, "It's really up to the school's discretion, except for the forty in the bench area." On occasion, Purdue will issue a sideline pass to a former player or corporate bigwig, expecting them not to crowd the bench area. Schott observes:

> If the school has corporate deals with companies, they may ask for sideline passes. We're pretty frugal with those but they do exist. Officials have the final say and if they think the visitors are getting too close to the sidelines, they'll push them back.

As long as participants in the game are not being harassed or distracted, the NCAA and NFL don't want to get involved in regulating the population flow on the sidelines. And even if the colleges don't like having to turn down entreaties for sideline passes, sometimes the alternative is worse. Case in point: Purdue. Schott remarks:

> For years we weren't very good in football so there wasn't much demand for sideline credentials. Now that we've gotten good, there are more requests.

Submitted by Rachel Rehmann of Palo Alto, California.

Why Do Crickets Chirp at Night? What Are They Up to During the Day? And Why Does It Seem That Crickets Chirp More in the Summer?

The answer seems obvious: Crickets chirp at night because that's when *we're* trying to sleep. But perhaps our application of Murphy's Law (the Imponderables Corollary: "All acts of nature can be explained by their ability to annoy us to the maximum extent") isn't what is uppermost in crickets' minds. Come to think of it, very little is likely to be uppermost (or bottommost) in crickets' minds.

Crickets chirp at night because that's when they are most active. Most cricket species—and there are about 100 just in North America—are nocturnal. They come out at night to find their two most pressing needs: food and crickets of the opposite sex.

During the day, crickets are relatively dormant, hiding from predators beneath rocks, in the grass or trees, or in soil crevices. By lying low when the sun shines, they are hoping to avoid confrontations with small owls, snakes, mice, frogs, raccoons, opossums, and other creatures that might try to hunt them for food.

Most entomologists believe that only male crickets chirp. They chirp by rubbing the two covers over their long wings together by using what is usually called the "scraper and file" technique. The cricket lifts one wing cover to a forty-five-degree angle (the scraper) and rubs the front end of it against the other wing cover (the file). Specialized veins in the wing covers make this possible: the file surface is rough while the scraper is relatively sharp. Crickets are "ambidextrous chirpers"—each wing can serve as both the scraper and file, and commonly crickets will switch off, presumably to prevent fatigue and excess wear on the file. The chirping sound will be the same regardless of which wing cover is used, but different species of crickets can be identified by slight differences in their "songs." The cricket is so famed for chirping that its name is of "echoic" origin (*cricket* is derived from the Old French *criquet,* an attempt to echo the chirping sound of the insect).

The primary purpose of chirping seems to be to attract female crickets. As Blake Newton, an entomologist at the University of Kentucky put it:

> Only males chirp, and they do so to attract females. This helps the females find the males. It is a big world out there for dating crickets!

Since each species creates a slightly different song, a female cricket of one species will not be attracted to a cricket from the wrong side of the tracks. Females are more active at night. Like their male counterparts, they are hiding from male predators during the day, and according to David Gray, biologist at California State University, Northridge, they are also busy laying most of their eggs in the daytime.

But sex isn't the only thing on a chirping cricket's mind. David Pickering, owner and webmaster of Chamowners Web (http://chamownersweb.tripod.com/), a site devoted to chameleons and crickets, wrote to *Imponderables*: "There are special songs for courtship, fighting, and sounding an alarm."

Chirping in the nighttime confers several other advantages to crickets regardless of what they are seeking when they sing, as Blake Newton elaborates:

> I can speculate that the special calmness of the night would allow the sounds of a chirping male cricket to emanate equally without distortion from the source. The chirps of the males not only attract females of the same species, but repel other males, thus resulting in distributing males in a way that would increase the mating success of females in the population.

Sound travels best when the air is calm, so nighttime is the right time for crickets. As anyone trying to sleep when crickets are chirping can attest, they can be quite loud. Cricket chirps have been known to travel over a mile in ideal conditions, but some crickets aren't content to leave their range to the vagaries of weather. In some cricket species, the wing itself acts as an amplifier; others burrow into long holes in the ground and chirp while inside, creating the kind of tunnel effect that echoes and augments the volume.

Entomologists used to believe that chirping was crickets' only way of communicating with one another, but we know now that they are capable of vocalizing. The sound generated is so high-pitched that humans are incapable of hearing it. The vocalizations seem to be some form of male bonding (female crickets don't seem to "talk"), perhaps a way of one male to tell another male of an impending predator.

Blake Newton mentions that the time of day or night that crickets chirp seems to be species-specific: Some species of crickets do chirp primarily when the sun shines. Captive crickets, such as those studied by entomologists in laboratories, are apt to chirp at any time, but with a definite bias toward the nighttime.

And yes, crickets do have a preference for hot weather. They *do* chirp more in the summer. When the weather gets cold, crickets not only stop chirping, they stop moving! Like other insects, crickets are cold-blooded. When the temperature rises, their metabolism increases, and the scraping of their wing covers is faster—so they chirp faster on hotter nights than cooler ones.

The relationship between chirping rate and temperature is so established that it is common folk wisdom to count chirps in lieu of consulting a thermometer. Just count the number of chirps from a cricket in fifteen seconds, and then add forty to that number. The sum is supposedly the current temperature in degrees Fahrenheit!

Submitted by Alexei Baboulevitch of Mountain View, California.

How Do They Put the Hole in the Needle of a Syringe?

Needles are used to poke patients. But are needles poked to create the holes through which the vaccine is pumped into our veins? The answer is a resounding no. As Jim Dickinson, president of K-Tube Corporation, wrote us:

> I have been involved with making the stainless-steel tubing used for hypodermic needles for the past thirty-four years, and the question about how the hole is put in this tube has been asked many times. The secret about the hole is that we don't put it in after, but before!

How? The answer comes from Michael A. DiBiasi, a senior mechanical engineer at medical supply giant Becton-Dickinson, who proudly asserts, "I am the guy who, among other things, puts the hole in the needle."

> The stainless steel "needle" part of the syringe is more commonly referred to as the "cannula," and the "hole" that has aroused your curiosity is called the "lumen." Cannulae are produced from large rolls of stainless steel strip stock. Depending

upon the size requirements of the finished product, which is dictated by its intended use, the strip stock could be about as wide and as thick as a piece of Wrigley's chewing gum, and may range down to about the width and thickness of one of the cutting blades in a disposable, twin-bladed razor.

The steel strip is drawn through a series of dies that gradually form the strip into a continuous tube. As the tube closes, the seam is welded shut and the finished tubing is rolled up onto a take-up reel. In this manner, the entire roll of flat steel is converted into a continuous roll of tubing. At this point, the tubing may be anywhere from about the diameter of a common wooden writing pencil, to about the diameter of an ink pen refill tube.

Next, the tubing is drawn through a series of tiny doughnut-shaped dies that further reduce its diameter while stretching the material, which thins the cross section of the tubing wall. Depending upon the desired target thinness of the cannula, and the physical properties required of the finished product, this process may or may not be accomplished using heat. In general, cannula tubing that is to be used for injecting liquids into the body may be produced with an outside diameter of about thirteen-thousandths of an inch, with a wall thickness of about three-thousandths or finer. Thus, the lumen may be as small as six- or seven-thousandths of an inch.

When all of the reduction processes are complete, the tubing is fed onto another take-up reel for transportation to one of several machines which cut the cannula stock into specific lengths for the next operation—point grinding [the point of the needle is chiseled or filed until the point is at its proper degree of sharpness].

As the stainless steel tube is pulled and lengthened by the dies, the dies create a bright, mirrorlike finish on the outside of the needle. The seam where the cylinder was welded together when the sheet metal was rolled into a tube all but disappears during this stretching and polishing process.

Even with changes in the production of needles, the holes prevail, as Jim Dickinson explains:

> The most recent technology uses a laser to weld a very thin stainless steel jacket around the hole, where in older processes electric welding required a thicker jacket. Once the hole has been jacketed, we then make it smaller and smaller by squeezing the jacket down around it.
>
> When we squeeze the hole it elongates, but try as we can, we have never been able to squeeze it completely out of the jacket. In other words, we have never been able to close the hole.

Submitted by Matt Lawson of Tempe, Arizona. Thanks also to Ray Kelleher of Spokane, Washington; and Gregory Medley of Tacoma, Washington.

Why Do FM Frequencies End in Odd Tenths?

All numbers are not created equal. Even numbers have cachet, while odd numbers are the black sheep of the integer family. And if there is a numerical caste system, fractions are at the lowest rung, always subject to being rounded off to the next whole number. Maybe this explains why more than ten Imponderables readers wrote to ask why U.S. FM frequencies end in odd fractions.

When the Federal Communications Commission moved FM radio to its current location in 1945, it placed the FM band between the television channel 6 (82 MHz through 88 MHz) and the Federal Aviation Administration frequencies (108 MHz through 136 MHz). Each station was allocated two-tenths of a megahertz (100 kHz on each side of its frequency) to avoid interference with adjoining station. The FM band was divided into 100 channels, starting at channel 200 (88.1) and ending at channel 300 (107.9).

Robert Greenberg, the late assistant chief of the FM branch, audio services division, of the Federal Communications Commission, wrote to *Imponderables*:

> Since each channel is 200 kHz wide, the center frequency could not start right on 88 MHz, because it would overlap into television channel six's spectrum and cause interference to channel six. Similarly, the same reason holds true at the high end of the FM band. To protect FAA frequencies starting above 108 MHz, the carrier frequency for the channel 300 would have to be below 108 MHz.

The irony is that the first channel below the FM band, channel 200, or 87.9, is rarely used because it is available only for use by low-power radio stations, and is assigned only if it doesn't conflict with an existing television channel six.

So all the radio frequencies were bumped up one-tenth to odd numbers to accommodate a small number of tiny stations with few listeners.

How odd.

Submitted by: Rick Deutsch of San Jose, California. Thanks also to Steve Thompson of La Crescenta, California; Josh Gibson of Silver Spring, Maryland; Susan Irias of parts unknown; Nadine Sheppard of Fairfield, California; Fred White of Mission Viejo, California; Anthony Bialy of Kenmore, New York; Gilles Dionne of Mechanic Falls, Maine; Robert Baumann of Secaucus, New Jersey; Doris Melnick of Rancho Palos Verdes, California; and many others.

Why Do Streets and Sidewalks Sometimes Glitter?

All that glitters can't be hocked at the local pawnshop. The shiny stuff we sometimes see on roadways and sidewalks isn't valuable, but it is variable—many components of concrete may glitter.

The most common ingredient among the glitterati is probably the minerals, such as quartz, that are found naturally in stones. The stones are crushed into a sandlike consistency and mixed with cement to form concrete or as part of the aggregate mixture in asphalt. Sometimes crushed glass is used as well and glass glitters mightily when exposed to light.

Because of the constant wear on road and sidewalk surfaces, the glitter effect tends to increase with time. As Thomas B. Dean, former executive director of the Transportation Research Board, wrote *Imponderables*:

> The fine aggregate/sand used in portland cement concrete is like a natural mirror; that is, it reflects light. In theory, all aggregate in concrete is completely coated with cement. However, the aggregate on the very top surface of the street or sidewalk will lose part of that coating due to weathering and vehicular or pedestrian traffic. Once exposed, the light from the sun, headlights, streetlights, or other sources bounce off the tiny surfaces of the aggregate, causing the streets and sidewalks to glitter.

Sometimes transportation engineers might actively seek out reflective surfaces on the roads they are designing. If so, they may add glass as a reliable and inexpensive solution to this need, says Jim Wright, of the New York State Department of Transportation. For aesthetic reasons, designers might want sidewalks to have a shiny surface, and may smooth down concrete with a rotary or blade to let the minerals in the sand strut their stuff on the surface.

More often, glass is included as a recycling measure. In fact, according to Billy Higgins of the American Association of State Highway and Transportation Officials, sometimes extremely non-glittery used tires are thrown into the aggregate mix as well, to put them to better use than as permanent residents at the local landfill.

Submitted by Sherry Steinfeld of East Rockaway, New York.

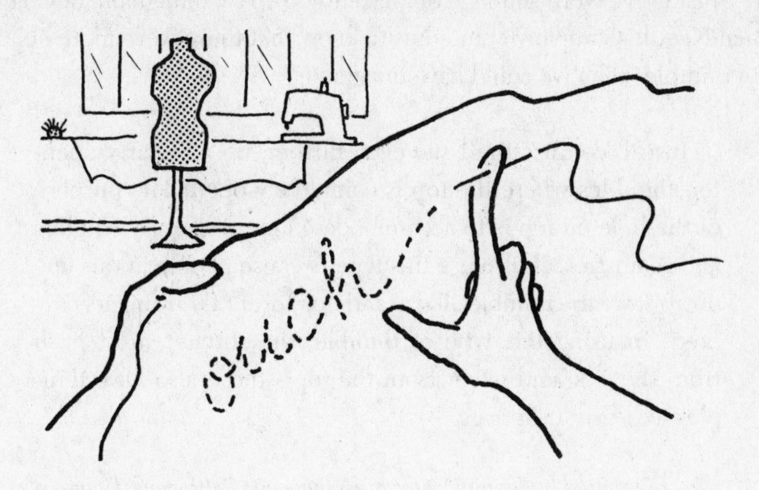

Why Do Thimbles Have Holes?

No, we're not referring to the big hole that sewers place their fingers into, but the little dimply indentations that digitabulists (thimble collectors) call "knurling." Were the little holes for ventilation (who wants a sweaty middle finger?), for decoration (we note that not all thimbles feature knurling), or to provide traction for the needle?

We spoke to a representative of thimble maker A. Meyers and Sons, who at first advanced still another theory that the holes were there to provide traction so that the stitcher could hold the fabric more securely. She checked with the sages at her venerable company and quickly changed her explanation: The holes were there to grip errant needles—they were a safety feature.

Other sewing experts we consulted agreed. As we heard from a representative from Silent Stitches Needlework:

> Those innocent little indentations on the top and sides of a thimble prevent the needle from slipping as you push the needle through the cloth being sewed.

Just as we were smiling complacently, Terry Collingham, of Colonial Needle Company, wanted us to know that there were more holes in thimbles than we could have imagined:

> Just to confuse the issue even further, we also carry open-top thimbles where the top is completely open. The purpose of the hole on top is to accommodate long fingernails. Tailors and seamstresses also use this type because people in this line of work wear a thimble all day and don't want their fingers covered. In using this type of thimble, the stitcher must push from the side and never from the top—these also have dimples to prevent slippage.

Submitted by Jenny Dennis of Hollywood, California. Thanks also to Bob Nissen of Syosset, New York.

How Do Birds Know Where to Peck for Worms?

When the red, red robin comes bob-bob-bobbin' along, it's not bobbing out of *joie de vivre*. The robin probably has one thing on its mind: food.

Different birds use different techniques to locate food. For example, sandpipers have long bills with sensitive tips—they might be able to feel worms that they can't even see. Kiwis' nostrils are located at the tip of their bills, so they can often smell their future prey in a way that other birds cannot.

But chances are that most of us are observing robins when we think about this Imponderable. Common garden birds in North America, robins exhibit striking routines when hunting worms in our backyards. Characteristically, a robin will cock its head at a funny angle just before pecking, elongating its body with its head as far off the ground as possible, and then violently plunging the bill into the soil. Earthworms are thought to comprise about 20 percent of an adult robin's diet, but virtually all of young nestling robins' nutrition, and parents must hunt and deliver their nestling's worms, too. Robins sometimes come up dry when pecking, but their "batting average" is remarkably high, and ornithologists have logged success rates as high as twenty earthworms captured per hour.

When you ponder the possibilities, it's conceivable that a bird could use any of the five senses to figure out where to peck:

1. Visual—maybe they can see the worm, or see movement on the ground that tips off the presence of worms.
2. Auditory—robins might be able to hear worms moving below.
3. Taste—a soil hors d'oeuvre might provide clues to the location of the worm.
4. Olfactory—the robin might be able to smell worms.
5. Vibrotactile—perhaps the robins' feet can pick up vibrations created by worms under the ground.

Much to our surprise and pleasure, biologists haven't ignored this Imponderable. Scientific foundations might not provide multimillion-dollar grants to study worm seeking, but nevertheless some scholars have conducted hard research into the feeding tactics of *Turdus migratorius* (robins don't have the most fortunate scientific name). Our first hero is Frank Heppner, Ph.D., who performed the earliest controlled experiments on this Imponderable in the mid-1960s. His biggest challenge in designing the study was trying to figure out how to isolate the different senses of his captured robins. Heppner made no assumptions about which of the five senses robins used to snare worms. Here's how he approached all five of the possibilities.

Smell—Heppner coated worms with rotten eggs, decaying meats, rancid butter, foul-smelling acids, and other putrid-smelling substances. This turned off the robins about as much as spraying perfume on a beautiful woman would turn off Don Juan. The robins ate the worms eagerly, not indicating any reaction to the foul smells. If the worms emit smells that provided olfactory clues to the robins, the birds still found and ate the prey even without the "good worm smell." (Robins, like most birds, have a poor sense of smell, so this sense was always suspect as the key factor in finding worms).

Touch—Heppner drilled wormholes in the ground and placed dead worms in the holes. The robins snapped up these worms, too. So if robins feel worm movement through their feet, they obviously wouldn't obtain any such advantage from dead worms.

Taste—Heppner found no indication that birds were picking up samples of dirt "on speculation" to find worms. Robins were too successful in finding worms on their first peck to consider taste as a major factor.

Hearing—Heppner tape-recorded the low-frequency sounds that burrowing earthworms make and then played the recordings back to the robins when actual worms were not present. The birds completely ignored the *Earthworms' Greatest Hits.*

Visual—Heppner drilled holes that looked exactly like wormholes, but did not place any worms, dead or alive, in the holes. Robins were uninterested in what was inside these holes.

Heppner concluded in his 1965 report that robins see their worm prey before pecking. In the experiments he conducted, robins found and ate worms whether or not the worms were dead or alive, and regardless of what the worms smelled like. Heppner's research stood as the definitive experiment on robins' worm feeding, and if you look at the popular literature on robins, you'll find that most of the literature assumes the validity of Heppner's conclusions.

But if Heppner was the early bird on this research, a pair of young biologists, Dr. Robert Montgomerie from Queen's University in Ontario, Canada, and Dr. Patrick Weatherhead from Carleton University in Minnesota, still thought they could capture the worm. They weren't convinced that robins found worms only through visual means, and in the mid-1990s, they decided to test the thesis. In their article, "How Robins Find Worms," published in the British scientific journal *Animal Behavior*, the biologists revealed their doubts about Heppner's conclusions:

> In an experimental study of robin foraging behaviour, Heppner (1965) concluded that American robins locate earthworms exclusively by visual clues. He based this conclusion on a series of experiments in which robins were able to find earthworms placed in holes in a lawn (but still visible) even in the presence of loud white noise (which would have obscured any auditory clues).
>
> Our own field observations of robins suggested to us that they might also use other sensory modes while searching for earthworms. When they cock their head, they appear to be listening, and we have watched robins successfully foraging on lawns where the grass was long enough to make earthworms difficult to see. We also watched a captive robin catch earthworms buried in soil where we could detect no visual clues that would reveal an earthworm's location. Thus it seemed to us that auditory, olfactory, or vibrotactile cues might be used in addition to visual cues when locating prey.

Unlike Heppner, Montgomerie and Weatherhead didn't bother to test taste as a major cue for robins, presumably because they discounted any chance of its importance. The duo captured some robins and placed them in an outdoor aviary in Ontario. They used "feeding trays" filled with soil, into which worms were placed in random but known locations. Usually, the trays were set on the ground so that the robins could "jump" on to the soil of the tray and hunt for worms. At other times, usually to test the robins' vibrotactile abilities, the trays were placed at a slight angle above the ground, so that the robins could peck into the soil without standing on the dirt.

The only other big change from Heppner's methodology is that for most of the experiments, Montgomerie and Weatherhead used mealworms (which are smaller than the earthworms that robins usually eat in "real life") as the bait. Needless to say, robins ate mealworms with gusto, too, and when offered earthworms instead by the biologists, robins didn't act differently from the way they did when pecking at mealworms.

Just as Heppner did thirty years before, Montgomerie and Weatherhead devised ingenious experiments to isolate particular senses:

Olfactory—The scientists placed two live mealworms and two freshly killed mealworms in the feeding trays. The odor from the live and dead worms were deemed to be very similar, yet the robins were much more successful at finding the live worms. The conclusion: smell was unlikely to be a significant factor in robins finding the worms.

Vibrotactile—The feeding trays were suspended and angled so that the birds didn't stand upon the soil, yet the birds were successful at finding their beloved worms. Vibrotactile cues didn't seem necessary for the robins.

Visual—A few live mealworms were buried in random locations on the feeding tray. The entire surface of the tray was then covered with a thin, but opaque piece of cardboard. Then more soil was piled on top of the cardboard. The robins were deprived of any visual cues, and the cardboard probably eliminated or at least drastically reduced any remaining vibrotactile or olfactory cues. The results: Although blocking visual access to the worms reduced their success rates, robins were still rather proficient at capturing their food, much more often than would be expected by chance—robins seemed to be capable of finding buried worms without any visual cues.

Auditory—The first three experiments led the experimenters to the belief that hearing might play an important if not dominant role in robins' finding worms, so the biologists designed a test to stifle auditory input while still allowing cues for all the other senses. They buried live worms randomly and then rigged a device to emit white noise (from the middle of the soil tray) that was sufficiently loud to mask out any sounds emanating from the movement of worms. The experimenters acknowledge in their report that the white noise might have affected some possible vibrotactile cues.

Still, the results were striking. When faced with the auditory distraction, the robins didn't strike as often, and when they did plunge into the soil, their success rate was higher than a random strike, but not as high as when the noise was absent. Montgomerie and Weatherhead concluded that the birds were having trouble locating the worms. They presumed that the robins' decent success rate probably meant that the white noise was not completely masking the sounds made by worms.

The biologists tried to entice the robins to strike by playing recordings of the sound of moving mealworms and piping the sound through the soil when no worms were actually present in the soil. But the robins didn't fall for the "worm and switch." Perhaps the sound quality wasn't an exact match; perhaps robins require more than one sensory cue to "go for it."

Montgomerie and Weatherhead concluded that all of the experiments indicated that robins could find worms when auditory cues are available and some or all of the other cues are removed. But by no means did they prove that hearing is the only sense that robins rely upon to catch worms.

At the end of their report, Montgomerie and Weatherhead attempt to reconcile their conclusions with Heppner's. They believed that Heppner demonstrated that robins could capture worms when they could see them, even when auditory and olfactory cues were missing, but "he didn't test whether the robins could use other senses in the absence of visual cues." They theorize that robins probably use a combination of cues, with visual and auditory cues probably playing the prominent role. Even if hearing is ultimately more important, they believe, robins will use all the visual cues they can amass. As Montgomerie wrote in a note to *Imponderables*: "Heppner showed that robins *could* use vision but not that they *couldn't* use hearing."

Of course, we offered Professor Heppner rebuttal space, and he graciously responded, despite it "being almost forty years since I worked on the robin business." In this e-mail, Heppner revealed that he blindfolded robins in his study, and

> Blindfolded birds just sat on the ground—they didn't hunt at all. Robins are perhaps not God's most intelligent creatures—perhaps they thought it was night.

Heppner was receptive to other theories, but didn't sound completely convinced that he was wrong, although he opens the door a crack:

> I am sticking to my original conclusion—based on the available evidence, robins *can* find worms by sight (and perhaps under certain circumstances, they can do it by sound).

Heppner concluded his note with a charming aside. After devising ingenious methods to test various hypotheses, and being challenged thirty years after the fact by other creative researchers, he concedes that science, and Imponderability, must yield to moral concerns:

> The problem with this question is that you can't really do the crucial experiment. If you surgically deafened robins, and they still found the worm, there's your answer (earplugs

wouldn't do it, because there could still be bone transmission of sound). But even though robins aren't particularly lovable birds, I don't want to irreversibly deafen a bunch of robins just to satisfy my curiosity—this is not finding the cure for cancer, after all—and evidently, neither does anyone else.

Submitted by Roy Dunten of Hermistown, Oregon.

What Letters Does Campbell's Include in Alphabet Soup in Countries That Don't Use Our Alphabet (e.g., Greece, Israel, Egypt)?

And what about France? Does Campbell's include an accent mark over the *e*?

The media relations representatives at Campbell's aren't exactly inundated with this question, but they researched it for us and graciously responded with the disappointing answer: Campbell's Alphabet Soup is sold only in North America. We urge reconsideration. Contemplate the potential of Campbell's Cyrillic Soup!

Submitted by David Faucheux of Lafayette, Louisiana.

Who Decides Where the Boundary Line Is Between Oceans? If You're on the Ocean, How Do You Know Where That Line Is?

Much to our shock, there really is a "who." The International Hydrographic Organization (IHO) is composed of about seventy member countries, exclusively nations that border an ocean (eat your heart out, Switzerland!). Part of their charter is to assure the greatest possible uniformity in nautical charts and documents, including determining the official, standardized ocean boundaries.

All of the oceans of the world are connected to one another—you could theoretically row from the Indian Ocean to the Arctic Ocean (but, boy, would your arms be tired). No one would dispute the borders of the oceans that hit a landmass, but what about the 71 percent of the earth that is covered by sea?

The IHO issues a publication, "Limits of Oceans and Seas," that determines exactly where these water borders are located, but is used more by researchers than sailors. Michel Huet, chief engineer at the International Hydrographic Bureau, the central office of the

IHO, wrote to *Imponderables* and quoted "Limits of Oceans and Seas":

> "The limits proposed . . . have been drawn up solely for the convenience of National Hydrographic Offices when compiling their Sailing Directions, Notices to Mariners, etc., so as to ensure that all such publications headed with the name of an ocean or sea will deal with the same area, and they are not to be regarded as representing the result of full geographic study; the bathymetric [depth measurements of the ocean floor] results of various oceanographic expeditions have, however, been taken into consideration so far as possible, and it is therefore hoped that these delimitations will also prove acceptable to oceanographers. These limits have no political significance whatsoever." Therefore, the boundaries are established by common usage and technical considerations as agreed to by the Member States of the IHO.

Essentially, a committee of maritime nations determines the borders and titles for the oceans.

How would the IHO decide on the border between the Atlantic and Pacific? A somewhat arbitrary man-drawn line was agreed upon that extends from Cape Horn, on the southern tip of South America, across the Drake Passage to Antarctica. A specific longitude was chosen, so the border goes exactly north-south from the cape to Antarctica.

Of course, there are no YOU ARE LEAVING THE PACIFIC OCEAN, WELCOME TO THE ATLANTIC OCEAN signs posted along the longitude. But a sailor with decent navigational equipment could determine which ocean he was in—likewise with the boundaries between other oceans.

Unlike the United Nations, most of the time the IHO does not become embroiled in political disputes, presumably because the precise location of the oceans' borders has no commercial or military implications. Disputes are not unheard of, though. For example, Korea and Japan recently tussled about the designation of the sea that divides their countries. Traditionally, the body of water has been called the Sea of Japan, but Korea wanted it changed to "East Sea."

Perhaps we were dozing during some of the year 2000 hoopla, but much to our surprise, the IHO was involved with a rather important event in that year—the debut of a new ocean. The southernmost parts of the Pacific, Atlantic, and Indian oceans (including all the water surrounding Antarctica), up to 60 degrees south, were dubbed the "Southern Ocean." The name was approved by a majority of the IHO and went into effect in 1999, with Australia among the dissenters. Why wasn't this a bigger deal than Y2K?

Submitted by Bonnie Wootten of Nanaimo, British Columbia.
Thanks also to Terry Garland, via the Internet.

Why Do Priests Wear Black?

Imponderables readers are asserting their spiritual side. At least you seem to be curious about superficial questions about Catholic priests and the clothes they wear, and that's good enough for us. Most readers assume that every vestment was adopted for its symbolic meaning, but in reality many of the clothes priests wear reflect the everyday dress of nonreligious folks nearly two millennia ago.

As John Dollison, author of the whimsical but solidly researched book *Pope-Pourri,* put it:

> Because they believed the second coming of Christ was imminent, [early Christians] didn't bother to formalize many aspects of their new religion. Clerical dress was no exception—nobody gave any thought to what priests should wear during Mass; they just wore the same clothes that laypeople did. . . .
>
> Fashions changed over time, but the priests didn't. They stuck with the same clothes they had always worn . . . until their garments became so different from what everyone else

was wearing that they were associated exclusively with religious life.

Not until the sixth century did the Church start to codify the dress of priests, and mandate that special garb be worn outside of the sanctuary. Even if most Catholics have no idea of the reasons for the uniforms, Dr. Brian Butler, of the U.S. Catholic Historical Society, feels: "The Church wants priests to be recognized easily by the laity. This is in the interest of both parties." You're unlikely to see priests in pastels soon.

Some priests started wearing black vestments in the early days of Christianity, as Father Kevin Vaillancourt, of the Society of Traditional Roman Catholics, explains:

> The practice of priests' wearing black originated in Rome centuries ago. Since the priesthood involves a renunciation of pleasures that the laity can practice, black was worn as a symbol of death—death to these desires, and death to slavish attachment to the fashions of the world. They were to concentrate solely on the service to God and others.

But by no means was there uniformity among priests in their garb until much later. Professor Marie Anne Mayeski, of the theology department of Loyola Marymount University in Los Angeles, points out that no specific color was required until after the Council of Trent (1545–1563), and that a response to the Reformation might have been partly responsible for the codification of clerical garb:

> Perhaps Catholic and Anglican clerics did not want to appear less sober and upright than their Puritan challengers.

There are exceptions to the generalization that priests wear black vestments. Higher-ranking priests put a little color in their garb. Cardinals' cassocks feature scarlet buttons, trim, and inside hems; bishops and other higher officials don amaranth; and chaplains to the Pope wear purple trim. During liturgical ceremonies, the cardinals wear all-scarlet cassocks, bishops wear purple, while parish priests wear black, although there are even exceptions here—a few dioceses, especially in warm-weather areas such as South America and Africa, allow priests to wear white cassocks, with trim indicating their rank.

Submitted by Doug Ebert, of San Bruno, California. Thanks also to Keith Cooper of Brooklyn, New York; Douglas Watkins Jr. of Hayward, California; and Tony Dreyer and William Morales Jr. of parts unknown.

Since Priests Wear Black, Why Does the Pope Wear White?

Blame it on Pius V, who assumed the papacy from 1566 until his death in 1572. For centuries before that, popes wore red. Why the change? Reverend Monsignor Dr. Alan F. Detscher, executive director of the Secretariat for the Liturgy, explains:

> Religious men who became bishops wore a cassock in the same color as the habit worn by their religious community. Pius V, being a [member of the] Dominican order, continued the practice of using the color of his religious habit, even after he was elected pope. The practice of the pope wearing white continued on after his papacy. On some occasions, the pope will wear a red cape over his white cassock, this a reminder that the more ancient papal color was not white, but red.

Like other religious traditions, what might have started as a personal predilection became codified to the point where now there are elaborate agreements about color codes—you'd think we were talking about battling VH-1 Divas who feared clashing outfits. For example, when the queen of England visits the Vatican, she wears black, as she is technically representing the Protestant Anglican church. But when the Pope visits her at Buckingham Palace, she can wear chartreuse if that's what she fancies.

Submitted by a caller on the Jim Eason Show, *KGO-AM, San Francisco, California.*

Why Does the Pope Wear a White Skullcap, and When Did This Custom Begin? How Does He Keep It Fastened? And Why at Other Times Does He Wear a Double-Pointed Hat?

We have our old friend Pope Pius V to "blame" for the pope wearing white, but he did not originate the use of the skullcap, properly called a zucchetto. Its use goes back to at least the thirteenth century. The zucchetto resembles the Jewish skullcap, the yarmulke, but its original purpose was quite different.

In the Middle Ages, when Catholic priests embraced celibacy, a ring of hair was removed from the top of their head, the tonsure. Churches and monasteries of this era weren't renowned for their creature comforts—the purpose of the zucchetto was to cover the "bald spot" in order for these often elderly men to retain heat in cold, drafty conditions. A cap that was sometimes used by clerics in the same era, the *camauro*, covered the ears and the whole back of the head, and was even more effective in staving off the cold. The tonsure was eliminated after the Second Vatican Council, but the headgear has lived on.

At no point has the zucchetto been worn exclusively by the pope, but since a proclamation by Pope Paul VI in 1968, only members of the hierarchy are required to wear the skullcap. You can tell the rank of a cleric without a scorecard—the color of the zucchetto is a tip-off.

Only the pope may wear white, with the exception of orders whose habits are white, such as the Norbertines and Dominicans.

Once again, Pope Pius V claimed the white color in honor of his Dominican order. Cardinals wear red zucchettos. Patriarchs, archbishops, and bishops sport a violetlike amaranth red zucchetto, and the "lower" deacons and priests wear black, although few priests wear zucchettos anymore.

How do clerics fasten the zucchetto on their heads? Evidently with some difficulty. According to Reverend Monsignor Dr. Alan F. Detscher,

> The zucchetto is not fastened on, but merely is set on the back of the head. It can fall off with movement and sometimes has to be adjusted in order to keep it in place.

The mitre, the double-pointed hat that the pope wears during ceremonial proceedings, is even older than the zucchetto—dating back to the tenth century—and can be worn by bishops and cardinals, as well as the pope. The hat features two cone-shaped peaks, divided by a piece of material that can fold together.

The mitre has undergone so many transformations in form that it is hardly recognizable as the original, simple cone-shaped hat that was worn by laypeople in Rome. At one point in the twelfth century, the mitre was shaped with the two points on the end, with a "valley" in between. But as John Dollison, author of *Pope-Pourri* notes, this created a problem:

> The points reminded people so much of the devil that they became known as horns . . . so the popes rotated their hats ninety degrees. They've worn them that way ever since.

Submitted by Peter Geran of Bethesda, Maryland. Thanks also to Jennifer Gaeth of Decatur, Illinois; and David Forsyth of Denver, Colorado.

Why Is There No Light in the Freezer Compartment?

This is a mystery dear to the hearts of *Imponderables* readers, and, we admit, to ours, too. Many is the time when we have pranced into the kitchen around midnight, opened the freezer, and ever mindful of the importance of nutrition, contemplate choosing a Ben & Jerry's dessert based on which contained the most calcium. If we had sufficient reading light before we took the stuff out of the freezer to know that one half-cup of the Chocolate Fudge Brownie frozen yogurt contained 20 percent of our daily calcium requirement, it might not have been necessary to eat ten half-cups of Cherry Garcia to make sure our calcium needs were met.

We posed this Imponderable to marketing and engineering types in the appliance industry, and their answers surprised us a bit. We assumed that a lightbulb would affect a freezer's ability to keep the freezer area cold enough, but since users rarely keep a freezer door open for long, the experts assured us that the energy usage involved in installing a light in the freezer would be negligible.

One of our sources, consulting engineer J. Benjamin Horvay, former chairman of the technical committee of the American Society of Heating, Refrigeration, and Air-Conditioning Engineers, suggests that not everyone is obsessed with reading nutritional panels in the dark as we are:

> It is a question of cost and the consumer's perception of value. For the most popular refrigerator configuration, the one with the freezer compartment on the top, the rationale is that because of its location, the freezer is more apt to be illuminated by the kitchen light than the fresh food compartment.
>
> Another factor is the relatively infrequent use of the freezer. Studies indicate that the typical customer opens the freezer door only once for every four fresh-food door openings.

Since we first received this Imponderable, more than a decade ago, lights in freezers *have* begun to appear more often. Perhaps the growing popularity of the side-by-side refrigerator-freezer, with more space devoted to the freezer, might account for some of the rise of the illuminated freezer compartment.

All the experts we consulted mentioned cost as the main obstacle to installing lights in freezers. Dick Stilwill, of the National Appliance Parts Suppliers Association, notes that lights tend to appear in high-end models; lower-priced models will lack a door switch, wiring harness, socket, and bulb:

> The consumer determines what features they are willing to pay for, and hence the various models in a line will go from high to low price-wise as the features decrease.

Product planners at the big appliance manufacturers decide which models will carry which features. Not unlike automobiles, appliance manufacturers can sometimes sell a refrigerator for much more money by spending only a little on "high-end" features. But as Ron Anderson, manager of advanced engineering in refrigeration at Amana told us, there are also expenses involved in bringing out a multitude of different models with only a few features differentiating them. Anderson says that the product planners will do an analysis of what competitors are offering at various price points; freezer lights would rarely be the make-or-break feature to a purchaser.

Still, the smaller manufacturers have to watch what the market leaders are offering. If volume leader General Electric, for example, offers lights in the freezers of its mid-priced refrigerators, the smaller players would have to consider offering them, too, or risk being at a competitive disadvantage.

Anderson mentioned Kano Analysis, a method of evaluating what is important to consumers, as a tool for evaluating the importance of features such as freezer lights. Developed by a Japanese engineer, Nariaki Kano, Kano Analysis argues that some features of a product (or service) satisfy customers in ways disproportionate to its functionality or cost to manufacture. The lowest and most basic level of quality to a consumer is a "dissatisfier." This is an attribute of a product that is so fundamental to the consumer that if it functions well, the customer doesn't necessarily feel any satisfaction. For example, it's unlikely that someone will rave about her refrigerator because "None of my food has been spoiled." Keeping the contents of the refrigerator cool is a basic attribute; if it doesn't perform, the consumer will be livid, but if it does perform adequately, it evokes no emotional response in the customer. The only emotion that a basic-level attribute is likely to generate will be negative—if the dissatisfier isn't met satisfactorily.

The second level in Kano Analysis involves "satisfiers," performance attributes that please the customer if fulfilled, and disappoint when not present. If a customer feels that the shelving in a refrigerator does not help keep food organized better than his last refrigerator, he might be dissatisfied; one that allows him to find his food more easily will satisfy him. It makes sense for manufacturers to focus on "satisfiers," as satisfiers often provide the basis for meaningful contrasts between products. All the performance attributes that auto manufacturers stress in trade magazines fall into this category.

But surprisingly, Kano found that a third level, which he called "delighters," can be crucial to the success of a product or service. These are attributes or features that are *not* essential to high quality and consumers don't even consciously think they want. But they provide the consumer with a sense of delight, especially because these are features that the customer didn't expect.

When folks explain why they appreciate an appliance, an automobile, or a meal at a restaurant, they tend to focus on "satisfiers," but Kano points out that "delighters" have a disproportionate effect on consumers' satisfaction levels, even when the cost is minimal to the business offering them. For example, the consumer might cite high fuel mileage as the reason for appreciating his new car, when he is secretly delighted by the vehicle's cup holders, which are big enough to carry a Big Gulp cup from 7-Eleven, or by the push buttons on the stereo system. The restaurant patron might cite the succulence of the sirloin when praising a restaurant, when he was really wooed by the waitress bringing an extra portion of home-fried potatoes, gratis.

Can a light in the freezer really *delight* a patron? If you itemized the options in a refrigerator-freezer, few customers would likely spring an extra fifty dollars for a bare lightbulb, but is one lightbulb enough to swing a decision in purchasing a thousand-dollar item? This is the kind of question that product planners at Maytag and GE must contemplate.

Submitted by Thomas Ciampaglia of Hackensack, New Jersey. Thanks also to Richie Edgar of Delmont, Pennsylvania; Barry Davis of Brooklyn, New York; Kristi Lingen, Nicole Fusaro, and Nick Tabia of Commack, New York; Bill Jelen of Akron, Ohio; Mike Rude of Irma, Wisconsin; Matt Savener of Wymore, Nebraska; Sarah Bresler of Bloomdale, Ohio; and Joseph Grabko of Harrisburg, Pennsylvania; and many others.

Why Don't Most Ovens and Refrigerators Have Thermometers?

As with the last Imponderable, money rears its ugly head. Most of us have plebeian controls on our appliances. Our ovens have temperature dials, of course, and when the oven reaches its appointed degree of heat, the oven clicks, or a light goes off. Our refrigerators have temperature controls, but they read from one to ten rather than in degrees.

But this is not so for the upper crust. If you want to spring $5,000 for a top-of-the-line Jenn-Air or Sub-Zero refrigerator, you can have precise temperature controls. We prefer to spend $4,000 less and be stuck with the 1–9/ "cold" to "coldest" controls on our humble GE.

Most of the appliance experts we spoke to thought putting in an expensive thermostat in an oven or refrigerator was much sillier than installing a lightbulb in the freezer. Dick Stilwill, of the National Appliance Parts Suppliers Association, observed:

> When you set an oven or refrigerator to the proper temperature, the unit will maintain that temperature until turned off. Adding a thermometer is a placebo to tell the individual that "Yes, my unit is at the temperature I have specified." On newer ovens, they even beep at you to tell you that the prescribed temperature has been reached. Oven thermometers serve the function of [soothing the owner who feels]: "I don't trust my thermostat."

More than a few bakers have good reason *not* to trust the accuracy of their thermostats, which is why most serious cooks own oven thermometers and instant-read thermometers to measure the internal temperature of food.

Unless there is an obvious malfunction, refrigerators are much less worrisome than ovens. Amana's Ron Anderson points out that a "looser" control works almost as well as a thermostat in a refrigerator, as temperatures vary within the unit anyway: "It would be misleading to track the temperature in just one spot."

While a ten-degree discrepancy in an oven might affect the results of a leg of lamb or a pastry, slight variations in temperature are unlikely to raise safety issues in a refrigerator or freezer. As Anderson puts it,

> There's enough thermal mass that the body of the food product will stay nearly the same temperature all the time. You might see short-term temperature swings in the refrigerator between the low thirties to forties. This really doesn't make any difference to the inside of a watermelon or the jar of pickles, because their average temperature is going to be right

where you want it. If you place the thermometer in the wrong spot, consumers might get nervous.

The consensus of our experts is that a more precise thermometer/thermostat is likely to be more of a "satisfier" than a "delighter." The cheap dial on the lower-priced refrigerator is a mechanical connection instead of the much more costly line-voltage thermostat necessary for more precise temperature control. Frugal consumers are unlikely to want to pay up hundreds of dollars for built-in thermometer/thermostats when they can go to the hardware stores and buy stand-alone thermometers for a few bucks.

Submitted by Warren Harris of Carmichael, California.

Do Skunks Think Skunks Stink?

Skunks can dish out a foul scent. But can they take it?

If, like us, most of our education about skunks comes from animated cartoons, you might be surprised to learn that skunks don't spray their noxious scent cavalierly. According to Skunks Scentral's counselor Nina Simone,

> Skunks only spray as a form of defense. It is the last action they will take when frightened. Each skunk has its own level of what degree of fear will trigger a spray. Some will stomp three times as a warning before "firing," which will give the "perpetrator" a chance to depart.

What exactly happens when a skunk sprays? We asked Jerry Dragoo, interim curator of mammals at the Museum of Southwestern Biology in Albuquerque, who is quite the mephitologist (an expert on bad smells):

> A skunk's scent glands are at the base of the tail on either side of the rectum. The glands are covered by a smooth muscle layer that is controlled by a direct nerve connection to the brain. The decision to spray is a conscious one. The smooth muscle makes a slight contraction to force the liquid through ducts connected to nipples just inside the anal sphincter, which is everted [turned inside out] to expose the nipples. The nipples can be aimed toward the target.
>
> When a skunk is being chased by a "predator" and is not exactly sure where the pursuer is located, the skunk, while running away will emit a cloud of spray in an atomized mist. The mist is light and takes a while to settle to the ground. A predator would run through this cloud and pick up the scent and usually stop pursuit. I call this the "shotgun approach."
>
> When the skunk is cornered or knows exactly where the predator is located, it emits the liquid in a stream that usually is directed toward the face. This intense spray will sting and temporarily blind the predator. I call this the ".357 Magnum approach."

Perhaps cartoons aren't far off the mark. Dragoo's description of the "shotgun approach" is not unlike Pépé le Pew's "cloud of stink bomb" method of foiling enemies.

But does the spray repulse other skunks? Our experts agree: "Yes." Simone mentioned that when other skunks smell a whiff, they become agitated. It is unclear whether this is a chemical reaction to the smell, or if it signals danger to them. She compared skunks' uneasy behavior when they smell other skunks' sprays to "a dog before an earthquake."

Considering that skunks don't like the smell of other skunks, it's surprising that they don't use their "weapon" more often during "intramural" battles. One skunk expert e-mailed us:

> Skunks actually don't like the smell of skunk, either, and unless one is accidentally in the line of fire, it would never get

sprayed by another skunk. It's kind of like a skunk pact that they won't spray each other.

If only humans were as accommodating!

But seriously, folks, we must delve into the seamier side of skunk behavior, for internecine spraying isn't that unusual. The most common perpetrators of skunk-on-skunk abuse are juveniles. Janis Grant, vice president of North Alabama Wildlife Rehabilitators, wrote us:

> The only situation in which I have observed skunks exchanging liquid insults has been when I have mixed different litters of young skunks together. They proceed to have a "fire-at-will stink-off" for about four to six minutes, including growls, chirps, and foot stomping, then gradually settle down to cohabitation. I can't say if this is to establish alpha status or just to make everybody smell the same, but none of them runs away from the encounters—they just spray a few times, retire to their corners, and let it go.

Dragoo notes that just as juveniles display the stomp, chirp, and spray behavior, sometimes a weaker skunk will spray a stronger young rival "if it feels it is being bullied." But they have been known to spray unknown adult skunks, too, "because adult males are known to kill young skunks."

Dragoo describes skunks' reaction to being sprayed as "the same behavior as other animals when they are sprayed":

> They will slide their face on the ground to attempt to wipe the odor off. They will also groom themselves [lick hands and rub face] to help remove the odor.

Are skunks, like humans, more tolerant of their own stench than others? According to Dragoo, skunks are not as egocentric:

The skunk can spray without getting a drop on itself. Skunks are actually clean-smelling animals. It is what they hit that smells, well, like "skunk."

If a skunk is in a situation where it would get its own spray on itself, the skunk's chances of survival are usually low. An animal hit by a car will often get the liquid on itself, but usually after death. If a skunk is caught by a predator and in the midst of a fight, it can get some of the liquid on itself. But in those situations the predator likely has already been sprayed and has not been deterred. The skunk will spray to defend itself even if it gets spray on itself.

The chemical composition of the spray is the same from one animal to another with some potential individual variation, but the "smelly" components are the same. Their own spray is as offensive as another's. The difference is that they are likely not to get their own spray near their face, whereas they would aim for the face of a "rival."

I have approached live, trapped animals and covered the trap with a plastic bag. This usually keeps the animal calm. However, on a few occasions, I have approached high-strung animals that spray multiple times at the bag. They are then covered by the same bag. They are still agitated when covered, but this may be a result of their already being wired.

On one occasion, I peeked under the bag and did observe the animal rubbing its face along the bottom of the trap as if it were trying to "get the odor off." Then it sprayed me . . . in the face.

Submitted by Robert Brown of Millerton, New York.

Why Do Ice Trays Function Better When Put in Their Designated Area in the Freezer? Or Do They?

Even low-end freezers have a spot designated for ice making, while expensive freezers feature automatic ice makers that promise everything but frozen margaritas. We've often wondered whether there is a reason for the location of the ice section of the freezer. Our experts say: Yes!

If heat rises, why are the ice compartments usually on the top? On most freezers, the evaporator, the metal tubing that converts a liquid refrigerant into cold vapor, is located on the top of the freezer. The ice trays then benefit from actual contact with the cooling source, and thus chill partly via conduction. The ice section of the freezer is usually located in the coldest possible spot.

A freezer is supposed to provide a cold temperature throughout the entire unit. But in practice, folks at home tend to jam freezer shelves with frozen waffles, ice cream, mysterious Ziploc bags, and leftover meat that will never be seen and certainly not eaten again. Dick Stilwill, of the National Appliance Parts Suppliers Association, implores us to unstuff our freezer:

> The airflow in a freezer gets restricted when you jam in all the items you can and still get the door closed. Folks want ice to be formed *now,* dammit! So [manufacturers] position the ice process where the airflow is unrestricted and will make ice the quickest.

What's so important about airflow in a freezer? When you place an ice tray with cold water in the freezer, the temperature of the water is obviously higher than the rest of the freezer. So as the water in the tray starts to freeze, there is some evaporation. This evaporation creates a thin layer of air just above the cubes that is slightly warmer than the rest of the freezer compartment. If the air in the freezer doesn't circulate, this little layer of warmth will slow down the progression of ice formation. The solution is to blow away the warmer air and bring in the ambient temperature of the freezer.

In the ice tray department, there is plenty of space for air circulation. But if you crammed the same-temperature water in the same ice tray and wedged it in between, say, your Weight Watchers frozen dinner and your gallon of Häagen Dazs chocolate chocolate chip, there is nowhere for the warm layer of air above the cold water to go. You didn't know it, but every time you fill your ice tray with water and place it in the designated area, you are proudly exhibiting the powers of convection, as physicist John Di Bartolo, of Polytechnic University in Brooklyn, New York, explains:

> How can the temperature that the water "feels" be lowered? Mix the air up. Allowing the air to flow distributes the heat energy emitted by the water evenly throughout the cabin so that the temperature of the air in the immediate vicinity of the water is lower.
>
> The rate at which heat leaves the water is proportional to the difference between the water's temperature and the surrounding air's temperature. Therefore, the lower the temperature of the surrounding air, the faster heat leaves the water and the faster the water reaches freezing-point temperature.
>
> I have a "quick freeze" option on my fridge, and it's creepy how fast it works. All it does is increase airflow across the ice cube trays.

A fan not only circulates the air but also increases evaporation in the water—if there is less water to chill, ice will be formed more quickly.

Barring serious malfunctions, ice trays will "work" anywhere in the freezer, but with the power of conduction and convection on your side, why not do the right thing?

Submitted by Kevin Bragdon of Houston, Texas.

Why Are Fraternities and Sororities and Most Secret Societies Named with Greek Letters?

The history of fraternities in the United States goes back before there *was* a United States—to colonial Williamsburg. We could understand it if social clubs were named after English letters, but why Greek, when few if any of these societies' members were of Greek heritage?

When you think of fraternities, three words spring to mind: Greek, secrecy, and beer. The first college social club for men, the Flat Hat Club, was formed in 1750, and embraced secrecy and beer, but left out the Greek. The Flat Hat Club was a secret society (with a name like that, who could blame them?), formed at the College of William and Mary in Williamsburg, Virginia. Members met at the Raleigh Tavern in Williamsburg, and although the club seems to have had no academic pretensions, its members were far from ne'er-do-wells—one Flat Hatter was a young Thomas Jefferson. The Flat Hat Club did have a secret badge and secret handshake.

Not long after, "literary societies" started popping up in several colleges. Literary societies ostensibly helped students with academics, but often were excuses for students to talk about issues that were proscribed in classes. Curricula tended to focus on the "classics" (for example, both Latin and Greek languages were required for all undergraduates at William and Mary), but many students wanted to discuss what they considered to be more pressing and immediate problems, such as the impending break with England. The Flat Hat Club seems to have died within two decades, but the P.D.A. Society at William and Mary seems to have taken its place, and its letters were taken from the Greek alphabet (Phi Delta Alpha, presumably).

John Heath, a young Greek scholar, was denied admission to the P.D.A. Club at William and Mary and, undeterred, decided to form a secret society of his own where scholarship, as well as social interaction, would be prized. On December 5, 1776, the first meeting of Phi Beta Kappa was held at the ubiquitous Raleigh Tavern. PBK started with a nucleus of Heath and four of his friends but, influenced by Free masonry, soon built chapters on other campuses (for example, Phi Beta Kappa's Yale chapter started in 1780, and Harvard's one year later). In the beginning, PBK's activities were secretive. This was because, like the literary societies, Heath and friends wanted to talk about issues that couldn't be discussed in the classroom, and also because secret rituals, handshakes, mottoes, and membership badges were ways to bond the members (and perhaps to feel superior to students denied access to the symbols of PBK).

Virtually all of today's Greek societies borrowed elements of Phi Beta Kappa. Why did PBK's founder choose its name? John Churchill, secretary of the Phi Beta Kappa Society, wrote to us:

> In an age when Greek and Latin were learned languages, it's quite natural that Greek and Latin should have been the naming languages. PBK actually had designations in both: the familiar Phi Beta Kappa, which stands for the phrase *philosophia biou kybernetes*, which means "love of learning, the guide of life," and *societas philosophia*, which is Latin for

"philosophical society." That's why there is the legend SP on the back of the PBK key.

The three Greek letters that formed PBK's name were the letters that formed the initials of its secret motto—a choice that would be echoed by virtually every long-lasting Greek organization, even if the phrase itself remains secret to outsiders.

As the decades rolled on, Phi Beta Kappa distinguished itself from other societies by focusing on academics, and in the 1830s, when much opposition was waged against secret societies, PBK abandoned its tradition of secrecy. William Morgan, a disgruntled Royal Arch Mason, threatened to expose all of the secrets of the Freemasons. Morgan disappeared and many believed that the Freemasons killed him. In this atmosphere, Phi Beta Kappa decided to voluntarily abandon its underground rituals and has become an academic honorary society—membership is earned by achievement rather than invitation.

The Kappa Alpha Society, formed at Union College, in Schenectady, New York, in 1825, is the oldest of the remaining secret social societies. Clearly patterned after Phi Beta Kappa, KA established a beachhead at Union College in 1817—indeed, KA's two founders *were* Phi Beta Kappas. *Imponderables* has been unable to wangle how KA chose its name, but we'll wager the society boasts a two-word Greek motto starting with K and A.

Delta Phi opened up for business in 1827, and is noteworthy for at least one reason. DP was the first secret society to call itself a fraternity. Most subsequent social societies also dubbed themselves fraternities, but in the 1830s, there was constant turmoil about whether new social societies should be secret or "open."

The original purpose of secrecy at the societies founded at William and Mary wasn't just to exclude non-members. Brothers wanted to avoid the censorship of discussion topics in classes, and to avoid outside interference by any officials at the school.

But many college students, especially during a climate of anti-

Masonry, felt that secret societies were unnecessary. Two fraternities, founded ten years apart, illustrate the tension within the Greek movement. Delta Upsilon, formed at Williams College in Williamstown, Massachusetts, in 1834, was created specifically to be an anti-secretive society. Its name was derived from the motto *Dikaia Upotheke* ("justice our foundation"). Members were chosen from the top 10 percent of their class.

In 1844, Delta Kappa Epsilon was formed at Yale, and from the beginning, the fraternity was cloaked in secrecy. DKE announced that it was seeking members who were "in equal proportions the gentleman, the scholar, and the jolly good fellow." DKE shied away from emphasizing academics, possibly because its founding members were rejected from scholastically oriented societies on campus. We will refrain from any jokes about Gerald Ford and Dan Quayle being two proud DKEs.

Through Web searches, we were able to learn that DKE has a motto that it discloses to the public *Kerothen Filoi Aie* ("friends from the heart forever") but also a secret one, *D K Chi Epi eye Kai Aie* ("right and equity"). Neophytes in the fraternity are warned: "This secret motto should never be uttered otherwise than in a whisper and then only in the presence of those known to be Dekes." Likewise, during induction ceremonies, neophytes are required to swear "never to reveal any part or parts of the activities upon which I now enter whether or not my initiation is successful."

Although different fraternities have adopted different rituals, the dominance of Greek culture continues. Joe Walt, historian of Sigma Alpha Epsilon, wrote to *Imponderables* explaining why:

> Much of what we incorporate into our names, our patron deities, and indeed our rituals was inspired by the powerful classical influence on education, and indeed much of educated society, during the eighteenth and nineteenth centuries.
>
> Most of it comes originally from the Greeks, but much of it uses Roman names and references. During the eighteenth

> century, English literature referred to the classical deities with their Roman names, for Latin had been preserved and used far more than had Greek during earlier centuries.
>
> Thus SAE adopted a Greek-letter name, as did virtually all fraternities, but it celebrates Minerva, rather than Athena, as its patron Goddess. . . . noble Leslie DeVotie, SAE's founder, who wrote the SAE ritual, was aiming to become a minister of the gospel and did go on to seminary after he graduated from the University of Alabama, and he knew Greek well.
>
> In any case, we—all the fraternities—have developed with a happy Greek-Roman-Christian ecumenism in our rituals and symbols.

And what about sororities? Secret female societies didn't even exist until two were founded at Wesleyan Female College in the 1850s (one of them later turned into Alpha Delta Pi), but then it wasn't like women were attending college in droves in the eighteenth century. No Greek letters were associated with female societies until 1870, and the first Greek women's group was Gamma Phi Beta—established at DePauw University in 1870.

Because of the long traditions of these societies, and their many rituals and vows of secrecy, it's hard to imagine how they'll ever move away from the Greek names. Perhaps this represents a marketing window for some new groups. What's wrong with a social group with an English name and motto? We think "Joe's Club" and "Sally's Society" have a certain ring to them.

Submitted by John Galt, via the Internet.

Why Are Charcoal Barbecues Usually Round and Gas Grills Usually Rectangular?

The kettle shape of the famous Weber Grill was initially more a matter of convenience than inspiration. George Stephen worked as a welder at the Weber Brothers Metal Works, and was frustrated by how often his grilling attempts on open braziers were foiled by wind, rain, blowing ashes, and flare-ups. By creating a deep barbecue, he helped protect food from these elements.

His job was welding metal spheres together to create buoys. According to "The Story of Weber" at www.webergrillrestaurant.com,

> It was in these very spheres that his idea took shape. He knew a rounded cooking bowl with a lid was the key to success. He added three legs to the bottom, a handle to the top, and took the oddity home.

As public relations representative, Donna Myers, president of the DHM Group, a public relations firm that represents many clients in the barbecue field, told us: "The round kettle was pretty easy to make with no seaming."

Stephen designed his first barbecue kettle in 1951, Weber Brothers Metal Works allowed him to stamp the kettles, and they attained success quickly. Most of our sources would concur with Bruce Bjorkman, director of marketing for Traeger Grills, about the reason why most charcoal grills ever since have been round:

> Probably the best answer I can give you is that most [charcoal grills] are round because people are knocking off the Weber charcoal grill, which was one of the first mass-produced charcoal grills in America. The first mass-produced grill was a brazier produced by the BBQ Company. It was a round, open grill . . . and goes back to the 1940s.

No one can accuse Traeger of following in the footsteps of George Stephen—it offers barbecues in the shape (and color) of a pig and a longhorn steer ("no bull!").

Not everyone jumped on the bandwagon, though. Many other manufacturers have and still do produce non-round charcoal grills. J. Richard Ethridge, president of Backyard Barbecues in Lake Forest, California, recalls that his company made large rectangular charcoal grills in the 1960s. But Ethridge has moved on to round barbecues with a difference—Backyard offers grills in the shape of a golf ball (perched on a tee) and an eight-ball nestled on a "cue" stand. Both of these models are available with your choice of fuel—propane, natural gas, or charcoal.

And the reverse is true as well. You can find round gas grills, such as the space-age model offered by Evo, a Beaverton, Oregon, company, which makes round gas grills with a flat, solid cooking surface. George Foreman's outdoor grill is a propane-powered round model that looks not unlike a Weber Grill.

Bruce Bjorkman believes that the domed top of round charcoal grills might aid in creating a "convection radiant dynamic," so that food cooks a little more evenly as heat is bouncing back in all directions. Donna Myers notes that after Weber's success, plenty of other non-round charcoal grills, especially square-covered cookers, became quite popular and performed well:

> I don't believe that the roundness and depth were ultimately essential. What was probably discovered was that a lid with any shape would do the job.

Myers notes that the rectangular form of gas grills was almost certainly a matter of economics: "I'm not sure whether gas grill manufacturers would tell you that it was the cost that led to that shape, but I'm quite sure that was the motivation."

We found one who was more than happy to share exactly this experience. J. Richard Ethridge points out that gas grills are more complicated to manufacture than charcoal grills:

> I think the manufacturing process pretty much dictated the shape of gas grills. . . . It is very difficult and expensive to manufacture a big round grill. Our grill is twenty-four inches in diameter and it takes a 650-ton press (1.3 million pounds of pressure) to stamp out that big a round grill. Metal (cold-rolled steel) will only "stretch" so far. There are not many factories in the U.S. or Asia that have a 650-ton or bigger press—they are very expensive.
>
> If you look closely at the Weber charcoal grill, you will see that it is, indeed, round if you look at it from the top. But if you look at it from the side, you'll see that the top is flat at the top, and the bottom is oval. It is not truly round-ball shaped. On the other hand, square box or rectangular grills are very easy to make in any size. It is much easier to bend straight edges on a large piece of metal than to make a box shape.

Ethridge pointed out other issues that make it less difficult to manufacture rectangular gas grills. For technical reasons, it is easier and cheaper to craft rectangular burners, and it is difficult to disperse heat evenly when you use a rectangular burner in a round grill. Most gas grills also have attached lids, while charcoal grills do not. While it is easy to manufacture a hinge for a rectangular grill with a flat back, Ethridge found when he first manufactured the 8-Ball and Golf Ball grills, that Backyard had to design a special hinge for the round grill so that it would lift up the lid first and then open. Even Weber, whose round kettles dominate the charcoal grill market, manufactures rectangular gas grills, presumably for economic reasons.

We were curious about whether Weber claims any advantage to the round shape of its charcoal grills, and were a bit stunned when our query was met by this response from the legal department:

> As Weber is a privately held company, our policy is not to provide any information regarding the federally protected shape of our kettle grill. Although interesting to others, we

consider the subject to be a trade secret, and highly confidential.

We didn't realize that the Weber's spherical form was a secret, but the guarded response is proof positive that in the barbecue world, it's the steak, and the sizzle, *and* the shape that matter.

Submitted by Jonathan McPherson of Richland, Washington.

Why Do We Rub Our Eyes When We're Tired?

If the first sign of wisdom is knowing what you don't know, then many of the physicians we contacted could put owls to shame. This is precisely the kind of Imponderable that we often have trouble answering. Patients don't run to doctors demanding that their eye rubbing be eliminated; doctors don't learn about eye rubbing in medical school—it's not a clinical problem. Researchers don't receive grants from the government to research eye rubbing. Somehow, it was reassuring to hear so many doctors respond honestly, with "I don't know." More than one offered a speculation with the proviso, "Please don't quote me on it."

Of course, rubbing the eyes isn't good for you. At best it does no harm. At worst, it can infect or damage your cornea. Adults know this, which might be one reason why babies and small children seem to rub their eyes more often than adults. The tendency of even the youngest babies to rub their eyes when tired indicates that eye rubbing is not a learned response.

Our medical experts were in three main camps:

RUBBING SLOWS DOWN YOUR METABOLISM AND HELPS YOU GET TO SLEEP

If you rub your eyes, you apply pressure to the rectus muscles that control eyeball movement, which in turn stimulates the vagus nerve, the long parasympathetic nerve that supplies motor and sensory fibers to much of your body. By stimulating the vagus nerve, you actually lower your heart rate and metabolism, making it easier to sleep. Electrical stimulation of the vagus nerve is used to treat some medical problems, particularly epilepsy and psychological disorders, in order to calm the body.

But when we're already sleepy, why do we need to rub our eyes to slow down our system even more? We're more sympathetic with this camp:

RUBBING IS AN ATTEMPT TO WAKE YOU UP

Most babies aren't exactly guilt-ridden about catching some shuteye, but sometimes try valiantly to stay awake, and this is one time when they're prone to rub their eyes. Are adults any different? Look at adults in a library, trying to study, and clearly headed toward torpor. Many of them take off their glasses and rub their eyes, valiantly (often futilely) trying to jerk their eyes, their brain, into refocusing. Dr. Arif Khan, associate professor of ophthalmology at Mount Sinai Medical Center in New York, concurs:

> Sorry, there is no "scientific" explanation for this phenomenon. I could say that the reason is to provide stimulation to the sleepy eyes to keep them from going to "sleep." But this is just an educated guess.

RUBBING HELPS KEEP CIRCADIAN RHYTHMS ON AN EVEN KEEL

"Circadian" refers to our daily cycles, and one of the most important "Circadian rhythms" is the ebb and flow of our metabolism that promote or discourage sleep. One part of the brain that controls our sleep-wake cycles is the "suprachiasmatic nucleus." One expert we consulted, Lenworth N. Johnson, a neuro-ophthalmologist at the Mason Eye Institute in Columbia, Missouri, offers a theory that might embrace and encompass the two above:

> Not everyone rubs the eyes when tired. Nonetheless, there are neurons (nerve cells) in the retina of our eyes that are involved in transmitting information on light-dark conditions. These are important in sleep-wake cycles, with projection of these nerve cells to the suprachiasmatic nucleus of the brain. I suspect rubbing the eyes may help to manipulate this signal.

In other words, the casual rub might be our attempt to micromanage our Circadian rhythms, as futile as that pursuit might be.

Submitted by Christopher T. Doody of Shortsville, New York. Thanks also to Wayne Good of Madison, Alabama; Aaron Burke of Saranac Lake, New York; and Ronit Amsel and Avi Jacobson of Montreal, Quebec.

Were Roman Chariots as Wobbly and Flimsy as Depicted in Movies? If So, How Could They Be Used Effectively in Wartime?

If the defense of Rome was dependent upon the stability of its chariots, no wonder the mighty empire fell. The Romans may have been decadent, but they weren't dumb: the chariot was far from a potent weapon of war.

Chariots were created almost two thousand years before the Roman Empire, and were suitable only for flat terrain. Earlier, ancient armies used them only because horses were not bred large and strong enough to withstand the weight of an armored soldier with weapons. The Roman legions rode on horseback, and reserved chariots for ceremonies and games such as the infamous Roman circus. According to military historian Art Ferrill, author of *The Origins of War,*

> Caesar's troops were amazed by the British use of them. For all practical purposes, chariots were last used in ancient Near Eastern warfare by the Assyrians.

What were the problems with using chariots for transportation? Let us count the ways. Carol Thomas, professor of history at the University of Washington, acknowledging research by J. G. Landels, *Engineering in the Ancient World,* which addressed these issues directly, wrote to us:

> The wheels of chariots probably were not flimsy. They were either spoked (in one of a couple possible arrangements) or solid. They may well have been wobbly, though, because any of the three or four possible arrangements for attaching the wooden wheels to the wooden axles of the vehicles led to grooves being worn on the bearing surfaces, which give the components more freedom of play.
>
> Acccording to Landels, there's no evidence of any swiveling mechanism in two- or four-wheeling vehicles, which would have allowed a chariot to corner with all wheels turning. Thus a driver had to skid at least one wheel around a corner, which also would have added to wear on the wheel-axle joints, reduced efficiency of the use of the animal-derived motive power, and also reduced stability of the vehicle.

Lest you think that these fits and starts would be handled with the finesse of a Lexus, think again. Stanley Burstein, professor of history at California State University and former president of the Association of Ancient Historians, wrote:

> My guess, for what it is worth, is that chariots were every bit as unstable as they appear, since they had no suspension systems and would react violently to any surface unevenness. They would also have been hard to control because ancient western horse harnessing techniques were extremely inefficient.

Abandoned as a war vehicle, the Romans adopted chariot racing as the ancient equivalent of a NASCAR race: the danger was half the fun.

Submitted by Gregg Cox of Wichita, Kansas. (For more information about the harnessing systems for Roman chariots, see http://www.humanist.de/rome/rts/index.html.)

Why Are New CDs Released on Tuesdays? Why Aren't New Books Released on a Particular Day?

Some things you can count on. Movies are released on Fridays. Diets start on Mondays. CDs are released on Tuesdays.

The Friday release of movies makes sense. A sizable majority of filmgoing occurs on the weekend, and studios can point toward a huge opening weekend by coordinating advertising and talk-show appearances by stars during the week. One of the reasons why Thursday night has become a battleground for young-skewing shows on the television networks is that movie studios spend huge bucks advertising their new films on that night to maximize attendance on the first weekend. The TV networks want to extract higher fees for those ads, which are based on the number of eyeballs tuning in.

Diets on Monday? The perfect time to work off the pounds you gained overindulging in food (tubs of popcorn at the movies?) and drink over the weekend. And self-sacrifice might as well coincide with the beginning of the dreaded school- or workweek.

But Tuesday seems like a colorless choice to launch new music (and videotapes and DVDs), especially when traffic in stores is highest on the weekends. Why was it picked? We had a theory, which was that the change occurred so that new releases would be given seven full days of sales history in order to attain the highest position possible on the *Billboard* charts, the bible of the music industry. But no less than the director of charts for *Billboard*, Geoff Mayfield, fingers another source:

> The culprit was not our charts, but the UPS man. As more and more chain stores received their new-release shipments directly from the labels' distributors, rather than from chain headquarters, stores at the end of a delivery route were at a competitive disadvantage to those which received their product earlier in the day on the dates when important titles came to market.

The uneven pattern of distribution occurred because UPS and other delivery services didn't provide service on Sunday, and the big chains were leaning on distributors to get new product as early as possible on Monday. All things being equal, the record labels would prefer a Monday launch, as Nielsen SoundScan, the company that measures record sales that form the basis of the *Billboard* charts, tracks sales from Monday through Sunday.

But four different sources, independently, used the expression "even playing field" to describe the relative fairness of Tuesdays for laying down new releases, and Tuesday seems to hit the "sweet spot" of providing maximum time for new recordings to hit the charts while satisfying the demands of retailers. Jim Parham, of Jive Records, elaborates:

> Most independent music stores buy from wholesalers called one-stops. The extra day, Monday, allows these wholesalers to ship to these accounts for the product to arrive on street date [i.e., Tuesday] or only one day prior. If street date were on a

> Monday, these stores would have to have the product delivered on the Friday before street date. When this happens, the label loses control of the release date, especially on stores not honoring street dates and selling the product early. This creates a chain reaction and you can lose a significant amount of sales that will not count toward the first-week chart position, as SoundScan sales are measured from Monday to Sunday.

Parham observes that the "street date" issue isn't as intense as it once was, as chain stores now dominate the market, and they tend to "jump the gun" less frequently than independents.

A uniform street date has other advantages. A source at Rhino Records, who preferred to remain anonymous, told *Imponderables* that the Tuesday street date allowed the production people at the label to set up systems that culminate in shipments every Friday that should hit the stores on Mondays. If there are delivery problems, a Tuesday launch schedule allows stores to resolve the issues on Monday. And letting consumers know that Tuesday is the day when new CDs are released is a way to drive traffic to retail stores during the week, according to Susan L'Ecoyer, director of communications at the National Association of Recording Merchandisers.

It must be tempting for stores to break the embargo and sell CDs that are lying around the stockroom. We were surprised to learn that the uniform laydown date is usually just a "gentleman's agreement." As Fred Bronson, *Billboard*'s "Chartbeat" columnist, told us, "I suppose if someone broke it consistently, suppliers could refuse to sell him any more records, which might be reason enough not to break the agreement." In practice, we couldn't find any evidence that any but a few scattered independent retailers were ever punished for selling product prematurely.

Many smaller music labels are quite content to have retailers stock the shelves as soon as product is delivered. For every CD that is launched with radio advertising, an in-person plug on "Total Request Live" on MTV, and a concert tour, there are many more independent label releases with no marketing budget and no prayer of ever making the *Billboard* charts.

There is little doubt that uniform laydowns work. Even if a Tuesday release loses one day of tracking by SoundScan, Geoff Mayfield observes that at the date he last wrote to us, July 9, 2003,

> We've already had fifteen albums *debut* at number one this year [in about six months], so albums obviously don't need a whole week to enter at number one.

By pointing all the marketing and advertising toward one day, free publicity can often be generated—the best recent example of this is not in music, but the book industry. When *Harry Potter and the Order of the Phoenix* was released at midnight on June 21, 2003, the publisher, Scholastic, attempted with great success to make the launch a media event: five million books were sold on the first day, numbers that exceeded the opening day's dollar grosses for the first two Harry Potter movies.

Most books are put on the shelves as soon as they are processed at the bookstore, as probably more than 95 percent of all book releases receive no marketing or advertising worth coordinating, but publishers will try to orchestrate the laydown of big books. We spoke to Mark Kohut, the national accounts manager at St. Martin's Press, who said his publisher's strategy is typical. St. Martin's generally will try a uniform release date with titles that have a chance to hit one of the major best-seller lists (particularly the *New York Times,* but also *Publishers Weekly, Wall Street Journal,* and *USA Today* lists), generally titles with first printings of at least 100,000 copies. St. Martin's releases its big titles on Tuesdays for exactly the same reason as the music labels. But the *New York Times* measures sales from Sunday to Saturday, so a Tuesday launch provides only five days of sales for the best-seller lists the first week. This might be the reason why Simon & Schuster chose a Monday laydown for Hillary Clinton's memoirs, which reportedly sold more than half a million copies on its first day of release.

The big specialty chains (e.g., Barnes & Noble, FYE) and megastores (e.g., Wal-Mart, Costco) are scooping up a greater share of music and book sales. Many of these retailers provide gigantic discounts and much better store placement for best-sellers. As a result, the pressure on record labels and book publishers to create instant best-sellers is more intense than ever. Although the day of release is a small part of the equation, it's a critical part.

Submitted by Allen Helm of Louisville, Kentucky. Thanks also to Scott Padulsky of Roselle Park, New Jersey; Christine Killius of Oakville, Ontario; Dave Frederick of Newark, Delaware; and Sam Bonham of Tellico Plains, Tennessee.

Why Are Loons Singled Out as Lunatics?

We suggest loons hire a new P.R. agency. While they sport a name synonymous with lunacy, they have much to crow about. For one, loons are the oldest living birds. We can trace their ancestry back about 70 million years, and there are loon fossils that date from 20 million years ago. Loons are among the most expert swimmers and divers in the avian world. They can dive down as low as 250 feet from the surface to catch their prey.

But despite their prowess in the water, loons tend to act strangely to our eyes on land. At least four typical behaviors of loons could be called "loony":

Loons are exceedingly awkward on land. Their legs are positioned far back on their bodies, which helps propel them when swimming and diving in the water, but makes it difficult for them to stand up. With their body weight saddled on legs designed for swimming, they fall forward when trying to walk. To move on land, they must crawl for short distances by pushing their legs and sliding on their breasts. Loons' nests are always at water's edge because of this lack of mobility on land.

Loons are awkward getting airborne from land or water. There are five different species of loons, and the heavier ones, such as the common loon, cannot alight vertically from a standstill on the water. To become airborne, they flap their wings rapidly while "running" or "patting" on the water to gain speed, not unlike an airplane taking off on a long runway. It is this patting, replete with noisy and manic flapping of wings, that if not "loony," is at least amusing to humans. Loons typically nest near large lakes to assure a long takeoff path. If a loon accidentally lands on solid ground, it's likely a goner.

Loons act funny. What's sexy to a loon isn't a turn-on to humans. To attract mates or fend off intruders, loons present several funny-looking courtship and territorial displays. For example, to fend off intruders, male loons rear up in the water, totally vertical, flapping their wings, while moving horizontally in this position.

Loons sound funny. The loon call is one of the most famous sounds of nature. Loons make at least five different kinds of calls, with names like "yodel," "wail," and "tremolo." In the literature about loons, you often hear sentiments like, "Once you hear it, you'll never forget it." Why are they so memorable? All three of these calls have been described as maniacal or eerie. The tremolo, usually employed to signal alarm, is sometimes called the laugh of a crazy person. The wail, used in social interactions, sounds eerily like a wolf's howl. The male uses a wild and manic yodel to scare off predators.

The bird experts we spoke to agree with Allison Wells, communications director of the Cornell Laboratory of Ornithology, that

The expression "crazy as a loon" derives from the calls made by the loons. The calls are often compared to the sounds a crazy person might make.

Of course, the weirdness of a sound is all in the beholder. To JoAnne C. Williams, state coordinator of the Michigan Loon Preservation Society, loon calls are music to her ears:

> The calls can sometimes be heard from quite a distance away, up to two miles. That's because sound travels well on cold days across the water. Many people think that some of the calls sound like crazy laughs, but I don't really think so. They are interesting and I really enjoy hearing them.

But if we conclude that "loony" is a reference to the strange calls of the bird, then etymologists might have a bone to pick with us. Check out a dictionary and you will see that most authorities agree that our name for the bird was derived from the Scandinavian word *lom,* meaning "clumsy," almost certainly a reference to loons' awkwardness on land. It's not hard to imagine Scandinavians, who emigrated en masse to the northern lake country of the United States, taking their name for the bird with them, as the loons they saw in the United States were exactly the same species of birds they saw in their home countries.

But how did a word for awkwardness get mixed up with *lunatic,* a word designating mental aberration? As it turns out, the derivations of the two words are not related at all! *Lunatic* comes from *luna,* the Latin word for "moon." Ancient people, through the time of the Romans, believed that overexposure to the light of the moon caused madness, or "lunacy." The word *lunatic* dates back to at least the thirteenth century.

To complicate matters even more, there is a Middle English word, *loun,* that means "madman" or "clown," that was the antecedent to the Scottish word *loon,* which refers to a simpleton or crazy person.

Shakespeare used *loon* in *Macbeth* ("The devil damn thee black, thou cream-faced loon!"), and Coleridge followed 200 years later in *Rime of the Ancient Mariner* (Hold off! Unhand me, grey-bear loon."). But there is no evidence that this word had anything to do with either "lunatic" or the bird. The British (and most other non-North Americans) call the bird "divers" rather than "loons."

"Loon" as the name of the bird stems from around the early 1600s, and the expression "crazy as a loon" developed among European settlers in North America. *Loony* wasn't coined until the late nineteenth century, and was slang for "lunatic." Most etymologists believe it was spelled "loony" rather than *luny* because of the already-established name for the bird and the expression, "crazy as a loon." Even though the derivations weren't the same, somehow the melding of moon madness, awkward birds on land, and eerie avian vocalizations melded into etymological confusion, if not lunacy.

Submitted by Patrick Brophy of Largo, Florida.

Why Are the Notre Dame Sports Teams Called "the Fighting Irish" When the School Was Founded by French Catholics?

When you conjure up an image of bruising football players, the French don't immediately spring to mind. But Notre Dame was indeed founded by a French Catholic, Father Edward Frederick Sorin, in 1842. Sorin had been a member of a religious order in France, Holy Cross Motherhouse of Notre Dame de Ste. Croix. This order specialized in missionary work and Sorin was chosen to lead a group of seven brothers to establish a center for Catholic education and missionary work in Indiana. Northern Indiana already had a strong French presence, as many of the first white men in the territory were French explorers, missionaries, and fur trappers of French-Canadian descent.

Sorin named his new school the University of Notre Dame du Lac, a tribute to his seminary back in France; "du Lac" was a nod to the two lakes on the forest land that Sorin had chosen to situate the university. While the university's original goal was to produce clergy, it soon welcomed non-Catholics and those interested in non-religious studies.

In the huge wave of immigration to the United States in the nineteenth century, many Irish Catholics settled in the Midwest; indeed, many Americans equated "Catholic" with "Irish." When Notre Dame started competing in intercollegiate athletics, many newspapers referred to its teams as the "Catholics," even though the school had no official nickname. In press accounts, many schools are referred to by their religious affiliation (yes, "the Catholics" battled "the Methodists" and "the Baptists" at football).

But where did "Fighting Irish" come from? What's a nice school established to train seminarians doing with a warlike nickname? Autumn Gill, a public relations representative from Notre Dame, told *Imponderables* that although no one knows for sure, there are two main theories (documented in a book Gill recommended, Murray Sperber's *Shake Down the Thunder*). The first theory is that "fighting Irish" was an epithet hurled *at* the Notre Dame team by fans of its opponent, Northwestern, in 1889. The Wildcat fans, who were behind in the game, yelled: "Kill the Fighting Irish, kill the Fighting Irish." The other story is that the term came from the lips of Notre Dame halfback, Pete Vaughn, who in a 1909 game against Michigan, tried to motivate his teammates (who were mostly Irish-American) when they were behind by yelling: "What's the matter with you guys? You're all Irish and you're not fighting." When the press heard about Vaughn's outburst, especially since Notre Dame went on to win the game, reporters dubbed the team the "Fighting Irish."

But the nickname didn't stick until the 1920s. In the first part of the twentieth century, Indiana press referred to the team as the "Catholics" or less flattering variations, such as the "Papists," "Horrible Hibernians," and even "Dirty Irish" or "Dumb Micks." Campus publications avoided the pejorative terms, and often referred to the teams by the school colors, "the Gold and Blue," and occasionally as "the Irish." Obviously, the campus administration wasn't wild about slurs against Catholics or ethnic groups, but the students embraced the "Irish" name and liked "Fighting" for its emphasis on spirit and playfulness. In campus publications, students insisted that "you don't have to be from Ireland to be Irish" and that naysayers should "cultivate some of that fighting Irish spirit." A late 1910 visit from Eamon De Valera, who was soon to be president of the Irish Republic, solidified the students' embrace of "Fighting Irish."

Three men popularized the nickname outside of South Bend. Knute Rockne, the legendary football coach, turned the Notre Dame team into a powerhouse. Rockne hired student press agents and encouraged them to use "Fighting Irish" in their dispatches. One of those press agents, Francis Wallace, moved to New York and became a successful sportswriter. He disliked the then-prevalent nicknames for Notre Dame, such as "Rambling Irish," "Rockne's Rovers," and "Wandering Irish," as all implied that the team's players traveled at the expense of their studies. Wallace's writings were picked up by the wire services, and he insisted on using "Fighting Irish." In 1927, President Matthew Walsh made it official, adopting "Fighting Irish" as the school's permanent nickname.

Of course, Catholics are more likely to root for Notre Dame than other religious groups, but Catholics from all over Europe and South America have emigrated to the United States, and yet seem loyal to a team named after one ethnic group. There were plenty of non-Irish members of Rockne's powerhouses, as the press loved to point out to him. But he always retorted:

> They're all Irish to me. They have the Irish spirit and that's all that counts.

Submitted by Jennifer Conrad of Springfield, Pennsylvania. Thanks also to Margaret Levin of Belle Vernon, Pennsylvania.

Why Did Bars Used to Put Sawdust on the Floor? Why Don't They Anymore?

Conjure up an image of a saloon in the Old West, and chances are you'll envision swinging doors leading to a long oak bar, liquor bottles on display, ornery hombres sidling up to the bartender, and sawdust on the floor. The sawdust wasn't there for visual effect, but for its functional utility. Christopher Halleron, bartender and beer columnist, explains:

> Remember back in elementary school when the janitor would use that sawdust-like substance to clean up the puke in the cafeteria? That's why they had it on the floor of bars. Puke, spilled beer, and all kinds of other foul substances end up on a barroom floor, and it's easier to clean up when it's absorbed by the sawdust and quickly swept away.

One of those other "foul substances" that the old saloons contended with was human saliva, according to an authority on all things liquor, author Gary Regan (http://www.ardentspirits.com):

> People used to spit all the time. Late-nineteenth-century bar books advise people looking for a job behind the bar not to spit during the interview, and also advise bartenders not to spit while on duty. I just read a "manners" book from 1934 that advises that people should not spit in company no matter what the circumstances, but then goes on to say something like, "but if you absolutely have to . . ."

In the Old West, one liquid that often was in short supply was water. Sawdust proved to be a remarkably useful substitute as a cleaning agent, as bartender Dan Morrison explains:

> I guess it dates back from times and places where water was scarce—so much so that actually washing a wood floor would be an extravagance. And if your floor was packed earth, it is a fact that sawdust soaks up the spills in a sweepable way, where dirt would crumble.

Dan himself has swept a mess or two away on a sawdust-laden surface and "marveled" at how easy it was sweeping up sawdust compared to scrubbing a solid surface.

"Baudtender," one of our online bartender correspondents, notes that sawdust was prevalent at an earlier time, and not just in commercial establishments:

> Sawdust is a sweeping compound—it wasn't just used in bars but in the finest homes. The natural resin in the sawdust adheres dust. You can still see higher-tech versions of this used today to soak up oil spills in auto garages.
>
> Remember that most of what gets spilled in bars is slippery when wet, and a sticky mess when dry. The sawdust served dual purposes—it gave traction over a spill to prevent slip-and-fall injuries, and it soaked up the gook so that it could be easily swept up later. But it wasn't just thrown all over the floor; it was kept in a bucket behind the bar and put onto spills as they occurred.

We remember fish stores and butcher shops spreading sawdust on their floors in the past, but why have they, along with bars, stopped the practice? The downscale image of sawdust-laden bars was a turnoff to many owners who didn't want to appear to be running establishments where patrons routinely tossed their cookies. (Ironically, sawdust was sometimes used to provide a certain "coolness quotient" to faux-downscale theme taverns later in the twentieth century.) Some bars served peanuts in the shells and encouraged patrons to toss the shells on the floor. The shells proved to be almost as absorbent as sawdust, albeit crunchier and more expensive.

Eventually, sawdust proved not to be an option even for bars that wanted to use it. Regulators honed in on food and drink establishments. Fire departments didn't like its flammability, especially when smoking was legal virtually everywhere. Health departments were testy about the notion that bacteria and insects were trapped in the sawdust along with booze and whatever spare bodily fluids that inebriated customers provided.

Today, where it is legal, a few bars provide sawdust floors for "atmosphere," and some mechanics' garages use synthetic sawdust to soak up grease and oil spills. Alas, bars contend with puke and spit with low-tech mops.

Submitted by Alicia Brooks of La Cañada, California.

Why Do Toads Have Warts?

Frogs are killing toads in the court of public opinion. Toads are plump; frogs are lean and streamlined. Toads hop about, but frogs leap. Toads spend most of their time confined to land, while frogs cruise around in water and resurface at their leisure. But worst of all, frogs have relatively smooth skin, while toads have bumpy "warts." Even worse, the myth persists that toads are poisonous and can give humans warts.

Toads don't even have warts (what we commonly call warts on toads are nothing but benign growths caused by viruses), so they can't pass them on to us. What we perceive as warts on toads are just places where the skin is thickened and cornified (covered by a cap of keratin, the type of hard tissue found in our fingernails).

But another type of "wart" is a thickened portion of epidermis surrounding the opening of the granular gland. These glands do contain poison—not enough to harm humans, although sufficient to sting the eyes or mouth. Rebecca A. Pyles, Ph.D., of the Herpetologists' League, says that in general, the more terrestrial the frog or toad, the more numerous are these tiny, individual poison glands. Toads also possess these glands in the big bumps located just behind each eye, above each ear. The purpose of these glands is clear, as Pyles explains:

> All amphibians have some poison glands in their skin, although the types and numbers of these glands differ among species. . . . Over 300 different kinds of toxins have been isolated from amphibian skin! One of the most poisonous vertebrates is a tree frog, which goes by the scientific name *Phyllobates terribilis*. One individual of this species, approximately one inch long, has enough toxin to kill about 20,000 white mice (20 gram in weight); in other words, one frog could kill a couple of humans.

Actually, all amphibians contain some amount of poisonous glands in their skins, and some toads have a relatively high concentration of them. If a predator scoops up a toad in its mouth, the poison burns the mucous membranes in its mouth—most attackers will drop the toad, and tend not to try to kill other toads in the future.

It's easy to understand how the "poison glands" can help a toad to survive, but what about those other "warts" that seem to exist only to make toads look ugly? We've seen three theories advanced to explain:

> 1. These bumps help break up the outline of the toad's body, thus allowing it to blend into the background environment, much like the patterns in camouflage fatigues are designed to blend in with particular terrains more effectively than a single color or shape.
>
> 2. Toads need water to keep their skin moist. When they are on dry land, they can easily become dehydrated. These bumps aid in the hydration of toads, as Pyles explains:

Moist skin means that the cells won't die, and also means that respiration (oxygen–carbon dioxide exchange) can occur across the skin—a particularly unique aspect of amphibians. The bumps themselves increase the amount of surface area for absorption of water, and more important, the channels between the bumps act as pathways that "pull" (by capillary action) water from the substrate over the flanks of the animal.

3. Do you think all toads look alike? Well, so do toads. In some species, males have a different number of bumps than females, so herpetologists figure that "warts" might help toads figure out who's who. Male toads aren't among the most discriminating lovers—they'll attempt to mate with virtually anything that moves. Not that there's anything wrong with that. But if they want tadpoles in their lives, it behooves them to find members of the opposite sex.

Submitted by Brandy Wright of Hanover, Pennsylvania.

Why Does Patting on the Back Induce Burping in Babies?

When babies get hungry, they want milk and they want it yesterday. Inevitably, overeager babies, especially those fed on the bottle, ingest air along with milk, and they experience the same gassy feeling as adults who ingest too much soda or beer at once.

There are only two ways for the baby to get rid of the air bubbles. Air can escape with the food from the stomach into the small intestine, but the passageway is closed right after a meal, nature's way of making sure we digest our food sufficiently before it rides through the gut. The second alternative is for the air to come back up, through the esophagus, back to the mouth. A valve at the entrance of the stomach tries to block food from coming back up (or else we'd be regurgitating more than frat boys on spring break), but if there is sufficient gas, the valve bursts open and baby emits a burp or what we hoity-toity people call an eructation.

Most adults have figured out methods to force a burp, but babies need a little help. Littleton, Colorado, pediatrician Don Schiff told *Imponderables* that gentle pats are usually sufficient to dislodge air bubbles that are trapped along the esophagus or in the stomach of the baby (and adults, too). Once the bubbles have been jolted free, the air rises and we are treated to that sound which is cute when babies do it but we get yelled at for emitting. (Some drakes even use burps as mating calls, not unlike their male human counterparts, albeit with much more luck.)

Back patting isn't just useful for humans, either. While researching this Imponderable, we stumbled upon a Web site that instructed how to burp a bottle-fed raccoon ("You must assist with burping by laying the baby across your lap and patting the upper back gently"). That begs the question of whether raccoon parents are giving their progeny such attention, although come to think of it, they are probably not giving their offspring bottled milk.

Submitted by Suzanne and Eric Thorson of Calgary, Alberta.
Thanks also to Steven Sadoway of Belmont, Massachusetts.

Why Are Thin-Cut Green Beans Called "French Style"?

"French-style" green beans are about as Gallic as Andy Griffith. The French don't eat American-style fat green beans, but favor haricots verts, a small-podded bean that is less than one-quarter inch in diameter and grows to approximately five or six inches in length.

Haricots verts are delicate, which means they must be picked by hand. The end result is a more elegant product, with a firmer, crunchier texture, and a nutty, more complex taste that only needs minimal cooking. ("You don't have to boil them into submission," notes Eddie Fizdale, owner of Peak Produce in Washington, D.C.)

Processed food suppliers wanted to cash in on the élan of haricots verts, so some enterprising marketing type created the idea of "French-style" green beans, and put them in cans and frozen food packages. Most "French-style" beans are merely ordinary American string beans halved in length and then sliced length-wise to resemble haricots verts. The result is often a tangle of stringy strands that aren't as bulbous as conventional green beans but can be messy-looking.

We spoke to Jim Kunkel, of C&W Frozen Foods in San Bruno,

California, who says that his company solved this appearance problem by choosing smaller, less mature green beans for their French-cut beans. C&W starts with shorter beans and simply cuts them in half.

"French-style" usually cost the same as "regular" green beans, a rare case in which something marketed as French is neither more expensive nor more prurient than its American counterpart. But haricots verts are much more expensive, so American farmers have taken notice. Haricots verts are now being grown in the United States and prices at the supermarket have inched downward as a result.

Submitted by Douglas Watkins, Jr. of Hayward, California.

Why Do Older People Tend to Snore More Than Younger People?

You know the drill. Right after the big family meal, the patriarch of the family hits the reclining chair and even before the television illuminates the den, the assembled throng hears the unmistakable sounds of "zzzzzzz, snnnnnnnort, zzzzzzzz, snnnnnnnort. . . ."

What causes snoring? The sound that we, the unluckily awake, hear is the vibration of air hitting the soft palate and uvula. Snoring is almost always caused by some blockage of airflow during breathing (since there is less room for the air to move because of the obstructions, the airflow is faster). Some of the most common obstructions include deviated septums, enlarged soft palates, uvulas, tonsils or tongues, and excessive tissue in the throat.

Of course, all of these obstructions are present when we are awake, too, but most people snore only when asleep because of the recumbent position and because the muscles supporting the organs in our throat are relaxed during slumber. ENT specialist Keith Holmes of Dubois, Wyoming, told *Imponderables* that as we age, the muscles in the mouth and throat "tend to become lax and flaccid," allowing the organs and tissues they support to protrude, blocking airflow. Elderly people also have higher incidences of deviated septums than younger folks, and also suffer more from obesity, as Steven C. Marks, M.D., professor of otolaryngology at Wayne State University, explains:

> Obesity leads to increased fat in the tissues around the throat, which causes the muscle tone to be decreased and the size of the airway to be decreased as well.

But Marks observes that snoring is found commonly in children, too. Maybe we don't notice it because no child has ever succeeded in wresting the Barcalounger from Grandpa.

Submitted by Chi Le of El Canon, California.

Why Do Many Whiplash Victims Feel OK the Day of the Accident and Much Worse Days Later?

Some folks experience little or no pain after being rear-ended, but then suffer greatly a day or two after. Maybe we're cynical, but we've always ascribed the delay to a quick consultation with a personal-injury lawyer. Orthopedists tell us otherwise. Depending upon the study, approximately 15 to 30 percent of patients examined experience neck pain soon after the accident, but that number balloons to 60 percent when evaluated later.

In the classic whiplash pattern, a stationary car is hit from the back by a fast-moving vehicle. Even if the victim is wearing a seatbelt, the body is thrust forward, and the head lags behind for a fraction of a second. Usually, the neck is bent backward, as if the victim were looking up at the roof of the car. The strain is exacerbated by the victim's state of relaxation (if the body could be braced for the crash, neck injuries would be minimized). The head then recoils, lurching forward into a hyperextension before swinging back to a neutral position.

Although most necks can withstand a forward thrust of fifty times the force of gravity, the rebound is often what causes the whiplash injury. If disks or vertebrae are ruptured, pain will likely occur right away, but with soft-tissue damage, pain is more likely to start from twelve to seventy-two hours later, usually centered in the neck, but often radiating to the shoulders and upper arms. When there is severe nerve or blood vessel damage, symptoms such as headaches and dizziness might occur, sometimes immediately, sometimes as long as months afterward.

Why the delay? Often, the victim exacerbates the injury unknowingly because there is no significant pain at first—a muscle strain caused by the jolt of the accident can turn into a spasm if the neck muscles are used too often or too harshly following the mishap. According to Berkeley, California, osteopath Richard O'Brien, if there is repeated use of the vertebra and ligaments around the neck after a whiplash accident, swelling will increase—it is the swelling that causes pain.

The accident often creates blood hemorrhages in the muscles and ligaments of the neck area. Although all of the mechanisms are not known, research indicates that inflammation usually increases in whiplash victims. Thomas A. Dorman, of the American Association of Orthopaedic Medicine, wrote in response to this question:

> It is characteristic of injured ligaments that the pain arrives some time after the injury. It is thought that this is due to the relatively sparse blood supply that increases only gradually after the trigger of the injury.

Type "whiplash lawsuits" into Google's search engine and you can see the controversy about the legitimacy of soft-tissue lawsuits. Whiplash is the perfect malady for the malingerer or downright fraudulent litigant, as many of the conditions that generate the symptoms are unverifiable by X rays or other "hard" diagnostic tools. Visit the Web sites of personal-injury lawyers, and you'll be apprised of the fact that there are scores of studies that conclude that whiplash victims who won legal judgments or settlements do not have a higher recovery rate from whiplash pain years after their accidents than those who have not sued. You'll find reports that indicate that juries are deeply suspicious of all soft-tissue accident claims, and that frivolous lawsuits are routinely thrown out of the courtroom. Go to a Web site representing the insurance industry, and you'll find studies indicating that frivolous whiplash suits are a financial drain on insurance companies (and ultimately, all consumers' premiums) and a waste of court time.

Believe us, we'd love nothing better than to make fun of personal-injury lawyers. It may be "convenient" for the symptoms of whiplash sufferers to lag hours or days after the accident, but we haven't found any convincing evidence that the delay isn't a legitimate medical phenomenon.

Submitted by Julie Hagaman of Ovideo, Florida.

Why Do White Styrofoam Picnic Coolers Have Blue Specks in Them?

We don't want to take a reader to task, especially one who poses an Imponderable we've wondered about ourselves, but we must amend one element of this question. Styrofoam is a registered brand of Dow Chemical, and Dow is evidently too lofty to deign to manufacture picnic coolers. So let's substitute the less elegant but more accurate "expanded polystyrene" (or EPS) for the trade name.

We spoke to Tom Conley, sales manager at Lifoam Leisure Products, the largest manufacturer of picnic coolers in the United States, who told us that when Lifoam gets raw polystyrene, it looks like salt granules—tiny, white particles. The polystyrene is then steamed, which makes the plastic expand into much larger "beads."

The EPS beads are white. So why the blue specks? Conley is proud to announce that their sole purpose is to look cool and to entice you to buy the cooler with the azure accents. Lifoam itself dyes the beads blue and mixes them with the unadulterated white beads to form the coolers you see at the local 7-Eleven or Wal-Mart. Not all picnic coolers have the blue specks, not even all of Lifoam's, but if you see one, chances are it's Lifoam's.

Submitted by Scott Walshon of Skokie, Illinois.

Do Fish Really Bite More When It Is Raining?

Some things we know are true: Where there is water, there are fish; where there are fish, there are fishermen; where there are fishermen, there are fishing stories; where there are fishing stories, there is disagreement.

We expected disagreement. What we did not expect was more theories than there are Commandments. We'll try to boil down and consolidate all the opinions we received, but we now realize one more thing: Where there are fishing theories, there are rarely short fishing theories.

On a few points, fishermen seem to agree. When fish are biting, it means that they are trying to find food for themselves. There are discernible patterns to when fish are most active in pursuit of food, related not just to hunger but climatic conditions in the water. And almost everyone agrees that rain seems to affect freshwater fishing, especially in shallow water, more than ocean fishing.

We posted this Imponderable on several online fishing forums, and received plenty of anecdotal evidence that fish bite more in rainy weather. "Jimbo's" response was typical:

> The fishing has always been good just before and during a rain, and that's the reason why so many of us are tempted at times to cast our better judgment aside and risk staying out

sometimes a little longer than we should with the approach of a storm.

In roughly descending order of popularity, here are the main beliefs about why fish bite more when it rains.

1. *Dinner Is Served!!*

Many fishermen echoed the sentiments of "Bazztex," an avid bass-fishing Texan:

> The primary reason rain makes fish feed is the food sources it exposes. Insects and small crustaceans, even small animals, get washed into the water. This sets up a food chain reaction with baitfish feeding on the bugs and bigger fish feeding on the bait fish attracted to the bugs.
>
> Heavy rains that cause a rise in lake or stream levels flood new cover. This exposes new food sources and attracts the fish that exploit the easy meals that await. The newly flooded landscape also gives the fish new cover to hide from predators. It's a win-win situation: Nature provides and fishermen enjoy the benefits.

Mark Bain, a fish biologist at Cornell University, confirms that the rain can even dig up new food opportunities for bottom-feeders, such as catfish:

> Catfish have many sensory organs on their bodies and they live in tough conditions along the bottom, where other fish would not be able to survive. Rain tends to stir up the water and disrupt the bottom. This helps the catfish when cruising for food, as they are able to sense new food sources opening up for them. Fishermen can take advantage of this by fishing for catfish in the rain, when the fish may be more aggressive.

2. *It's the Barometer, Baby!*

Before a rain, the barometric pressure falls. Fishermen believe that fish can sense the barometric change and get more aggressive. Captain John Leech, a full-time professional fisherman and bass guide, wrote to *Imponderables*:

> The study of weather will give us a bigger piece of the puzzle of fish behavior than any other single study. . . . After three days of any constant weather, the fish will start to become accustomed to the conditions and return to a normal activity. The passing of fronts is the change factor. Warm fronts are the fish-catching fronts. Cloudy weather, dropping barometer, south to west winds are the predominant conditions. Resident fish will move out from under the heavy cover to the edge and feed. The deep open-water fish will move to all breaks, even to the shallows to feed.

What explains this behavior? Professor Bain confirms that fish can sense barometric pressure changes, and the most likely explanation for this gift is to allow them to sense when food might be difficult for them to acquire (such as when there is a storm). Instinctively, then, fish may sense a drop in barometric pressure as a time to start eating "while the getting is good." Biologists don't know for sure exactly how fish sense barometric changes, but one common theory is that the "lateral line," a collection of hairlike structures along the flanks of most fish, is responsible. We know that fish use the lateral line to detect pressure waves from other fish to protect themselves when they are about to be attacked, even if they can't see or smell the potential predator.

At least one credible source minimizes the importance of barometric changes in affecting fish feeding behavior. B. C.

Roemer, president of ScentHead, a company that manufactures artificial baits, writes:

> Can a fish notice this small change and has it anything to do with feeding (the bite)? I don't know, nor does anyone else. I do know that just before a storm (and even in it) fish turn the bite on. This is a well-proven fact. But does a low barometer affect the fish's body to trigger the bite? I don't think so, assuming a bass is about at a 3-foot depth. Under normal swimming it would have to stay exactly at that level or the pressure from the water would increase or decrease a lot more than the small air-pressure change. So it's reasonable to disregard barometer readings. Something else is going on to produce the bite.

Roemer notes that a fish swimming even a few inches up toward the surface or lower toward the floor will feel a much greater change in pressure from their altitude than because of barometric fluctuations due to weather patterns. Mark Bain acknowledges the truth of Roemer's assertion, but adds that fish may be able to sense the outside pressure changes independently, in a way that we do not understand.

3. *The Eyes Have It!*

The clearer the water is, the more fish act defensively. In his article in *Field & Stream,* Jack Kulpa notes this effect:

> Even the biggest largemouth bass feels exposed and vulnerable in direct sunlight. On the brightest days these fish burrow into weeds or head for deep, dark water where they are all but inactive and unapproachable. Yet as daylight fades with the approach of a storm, even big bass are lulled into a sense of security. When that happens, they may strike suddenly and unexpectedly.

Mark Bain notes that rain tends to reduce water clarity. Although turbid water makes it harder for predators to attack, it also decreases a fish's visual access to food. Many experts believe that fish learn when the water turns cloudy that they had better look for food fast, before a storm renders it too difficult to find and eat a proper meal. For the fisherman, this can be a mixed blessing. Says Bain:

> If the fish turn off from feeding, that will be bad. But if they're hungry, they will tend to take bait more freely if it is presented directly in front of them.

With rain comes clouds, and when there is a cloud cover, less ultraviolet light penetrates the water. Fish are sensitive to light and are more apt to feed when there is relatively less light in the water. This is probably a major reason why the presumed best times for fishing are early morning and late evening, when temperatures are cool—these are both times of reduced light above and below water. When it rains and the cloud cover darkens the sky and the water, the fish may be tricked into thinking that it is actually late evening. This theory is hard to prove—we were unable to get a fish to comment on or off the record.

4. *It's the Water!*

A light rain aerates the water, which has the effect of naturally oxygenating the water in the same way that those little bubble machines do in an aquarium. Jack Kulpa notes that the combination of cooler water temperature and increased oxygen seems to give bass (and other fish) a burst of energy, sort of the fish equivalent of a cup of joe.

In his book *Keeper of the Stream,* author Frank Sawyer notes that the water seems to come alive after rainfall, partly because flies hatch in profusion, possibly because of the aeration. If flies are hatching, fish are trying to eat them, and fisherman are trying to capitalize on their prey's increased biting.

5. *Hear No Evil!*

Water is an excellent conductor of sound, so any noise generated will travel through the water. Johnny Hickman, an avid fisherman based in West Texas, shared a theory with *Imponderables* that posits that audio might be a key component in the answer:

> In a steady rain, the thousands of raindrops hitting the water will be pretty noisy underwater, providing a kind of "white noise" that will tend to hide man-made noises. Add to that the decreased visibility due to cloud cover and the constantly disturbed surface of the water and you get a situation that makes the fish less spooky.

6. *It's the Humans, Stupid!*

Could the human psyche play a role in success in fishing? One authority, Lesley Crawford, writes in his book, *The Trout Fisher's Handbook*,

> I haven't come up with enough evidence to convince myself that fishing is better in the rain. One thing that I do know is the importance of confidence. If you think and are confident that fishing is better in the rain, then you will catch more fish because you expect to, and, as a result, probably fish better, too—which is, perhaps, the real reason that you are catching fish!

But more than positive thinking might be involved. If it rains, less hardy (or less crazy) fishermen retreat to their campsites or automobiles (or nearby tavern) and there is less competition to catch fish, accounting for more success per active fisher. The fish may be more likely to feed when fewer humans are afoot and fewer boats are mauling the serenity of the water.

Fishermen, with luck, are a little brighter than the average trout, and anglers have gained knowledge about the predictable habits of fish during rain. If fish are known to retreat to an isolated inlet when it rains, hardcore fishermen will brave the elements to drop lines there. If they know that a storm will cause fish to withdraw to the lower depths, then fishermen will cast lower than they normally would.

THE DISSENTERS

A minority, but a vocal one, isn't so sure that fish do bite more when it rains, at least not consistently so. One group, the Forest Preserve of Cook County (Illinois), tabulated daily records of catches over a twelve-year period, from 1932 to 1943, with over 15,000 pounds of hook-and-line fish caught by its members.

The group's conclusion? There was a slight increase in catch rate when it rained at least one-half inch, but not at all on the days after. Surprisingly, bass bit almost twice as much when the water cleared up two or three weeks later, but it didn't seem to matter whether the weather was fair or cloudy, from what direction the wind was coming, whether it was day or night, or whether the barometer was high or low, rising or falling. The group also found no significant difference between the catch rate of men versus women:

> During the entire twelve years, men averaged 3.25 pounds per day, while the women averaged 3.22 pounds. Of course, the men say they hooked a lot of big ones that got away.

Others who have conducted more limited experiments have issued conflicting reports about the role of barometric pressure, cloud cover, temperature, and rainfall on biting patterns of fish.

All these different theories make our head spin, and since we can't seem to catch fish in any environment other than a stocked pond, we identify more with the sentiments of "Mudcat," a fishing hobbyist who preferred to theorize about the effects of rain upon humans, or at least one human:

> I know that rain sure makes me hungry. Just last weekend [during a rainstorm], I had eight tacos and two burritos in one sitting.

Submitted by Professor Elizabeth Goldsmith of Tallahassee, Florida.

Why Are We Instructed to "Remove Card Quickly" When We Swipe Our Credit Cards at the Gas Pump or Grocery Store?

Your credit card face is full of all sorts of information—your name, your credit card number, the expiration date, the snazzy graphics, the name and address of the issuing bank, and the logo of the credit card company. But the machine that swipes your credit card cares not a whit about any of that stuff. All it lusts after is the information held in that thin, horizontal black stripe that runs across the back of the credit card. That stripe, known as a "magstripe" (short for "magnetic stripe") contains tiny magnetic particles that can be magnetized so they each lie in one of two directions. These particles provide all the information that the bank, the oil corporation, or credit card company needs to haunt your next statement.

Just as the binary data on a computer, ultimately, is a series of zeros and ones, so are these particles magnetized to be oriented on the magstripe. These little iron-based magnetic particles are only twenty-millionths of an inch long. Once the province of credit cards, magstripes can be found not only on ATM cards, but student identification cards, library cards, and office-machine user ID cards.

The magstripe works on the basic principles of electromagnetism. Whenever a magnet moves, it generates an electrical field. If there is a metal wire (or other item that can conduct electricity) near the moving magnet, the motion will cause an electrical current to flow in the wire or other conductor. (The converse is also true: If you have an electric current flowing in a wire, the current will generate a magnetic field near the wire.)

When you swipe the credit card by sliding it in and out at the fuel pump, the movement of that magnetic stripe across the "read head" (the part of the card-reading device that interprets the data held in the particles) creates a tiny electrical pulse. The read head is capable of distinguishing between pulses in the particle magnetized to represent a one, or the other particle magnetized in the other direction to represent a zero.

The faster a magnet moves by a wire, the stronger the electrical current in the wire will be. This is the principle by which electric generators work. The faster you spin a magnet, the stronger the electric current you generate will be. The corollary is also true—the higher the voltage of your initial current, the stronger the magnetic field you will generate around the wire.

Credit card swipes are no different. When you swipe your credit card, you generate an electrical pulse in the read head. The faster you swipe the card, the stronger the electrical charge generated by each magnetized particle will be. You are encouraged to swipe quickly so that the read head can receive the strongest possible signal. The stronger the signal, the better chance the read head has of interpreting the data correctly.

Eventually, you may see these signs disappear. Larry Meyers, director of engineering for MagTek Inc., a Carson, California, company that manufactures card readers and specializes in "magnetic stripe card solutions," wrote *Imponderables* warning that these Remove Card Quickly signs might go the way of the dodo:

From a practical standpoint, most card readers today use electronic designs that feature AGC (automatic gain control). This allows the electronics in the card reader to automatically compensate for low electrical signals which occur during slow swipes. Thus, to a large extent, the need to "swipe quickly" has been eliminated.

But there's another reason why swiping quickly might still be a good policy for the prudent customer: Fast swipes are smooth swipes! A smooth swipe provides fewer read errors, primarily because card readers work best when the card is withdrawn in a continuous motion. Stewart Montgomery, a customer-service representative at MagTek, notes that the Swipe Quickly sign

> is to prevent the cautious person from moving the card at an extremely slow and uncertain rate. Moderate speed is best for the magnetic sensors used for credit-card magnetic-stripe reading. The typical acceptable speed is usually specified in a range between three and 50 inches per second.

That's quite a range. We don't think we've ever been so enthusiastic about paying, even by plastic, that we've managed the fifty-inches-per-second swipe.

Montgomery's point was echoed by Dave Lewis, a technical-support representative from Corby Industries, who argued that "fast" really means "steady":

> Magnetic-stripe cards hold a string of information usually defined on track two of the card. This track two is in a format set by the American Banking Association. There are generally as many as sixteen distinct characters within this track, all numeric or numeric equivalent. This is why there are sixteen digits on your credit card or ATM card.
>
> Since this numeric encoding is in a string, it is more likely to

be read correctly if pulled through the reading device in a uniform fashion. Swiping the card too slowly may cause space between the character information, causing a misread of the card. It's like pulling a train with an engine. If all the cars stay connected to the engine it is much easier to reach the destination quickly. A break between the cars will create space and slow down the train, especially if the cars need to be reconnected.

This is why people are told to swipe their card "quickly." It really means "steadily." People understand the term "quickly" easier. This is why the money machines ask for a quick swipe.

Speaking of swiping quickly, when we use our ATMs, why do banks remove money from our account right away? When we deposit our meager paychecks, it seems to take weeks for our money to be credited to our account. As long as the banks set the rules, perhaps this isn't an Imponderable after all.

Submitted by Amber Burns of Salem, Oregon.

Some Credit Card Reading Machines Ask You to Dip the Card and Pull It Out Rather Than Swipe. Does the Reader Pick Up the Data on the Way In or the Way Out?

The machines that ask you to dip your credit card are, appropriately enough, called "DIP readers." Once again, the industry has to adjust for those users who just aren't smooth enough. Card-reader manufacturer MagTek's director of engineering, Larry Meyers, explains:

DIP readers normally have a mechanical or optical sensor at the rear of the reader, which detects when the card is fully inserted. Using this sensor as a reference, the DIP card readers are then "programmed" to wait until the card is fully inserted, and then will only read card data as the card is being withdrawn from the reader.

This approach is taken because users often have an awkward time inserting the card into the reader. If the card is "jerkily" inserted into the reader, it is not an ideal condition to try and read card data. However, once the card is fully inserted, it is very easy for the user to withdraw the card in a smooth, controlled motion that optimizes card reading conditions.

Submitted by Ernie Capobianco of Dallas, Texas.

Why Are Flour Tortillas Larger Than Corn Tortillas?

Which came first: the big flour tortilla or the big burrito? If corn tortillas were as large as flour ones, and were stuffed accordingly, we'd be ordering fewer tacos at a pop, wouldn't we?

Although there are no strict standards, most corn tortillas are about six inches in diameter (although 5 ½ inches is another popular size for prepackaged corn tortillas) and rarely exceed seven inches. Flour tortillas can be found that are as big as a medium pizza (twelve inches), although smaller ones also are available that are no bigger than their corn counterparts.

Corn tortillas are thought to date back several millennia before Christ and became the staple of the Aztecs' diet by the sixteenth century. When the Spanish conquerors came to the highlands of Mexico, they found the native women pounding corn to make bread; the Spanish dubbed the bread *tortilla,* or "small cake." The Spanish later introduced wheat to Latin America, and flour tortillas, created only in the past few centuries, were the result.

Traditionally, the first step in preparing corn tortillas was soaking corn in a lime solution that eliminates the husks. The corn kernels are then ground with a stone into masa, and the dough is then combined with water and mixed by hand into a golf-ball-sized sphere. The masa is then rubbed between the cook's hands and patted into a flat pancake of approximately six inches. The six-inch dimension wasn't a function of any particular standard, rather it was as big a dough as could be molded easily by hand. If it were any bigger, the tortilla would flop over the outsides of the baker's hands and fall apart. The eponymous Maria Vega of Maria's Tortillas in Los Angeles, California, elaborates:

> The dough just doesn't hold its consistency in larger sizes. If you try to slowly and carefully pat out a larger corn tortilla, even on a table or other flat surface, it tends to fall apart and develop cracks. You can sometimes do it if you're very careful and take your time, but it's difficult.

Of course, commercial plants don't make corn tortillas by hand. Tortilla presses are capable of flattening a ball of masa in a second or two, making perfect six-inch circles of corn, and are even available for home use.

The dough of flour tortillas has a different consistency from corn—it is too soft and sticky to be patted between the hands or even to be subjected to the tortilla press (although we heard tales of a few skilled Mexican women who have made paper-thin flour tortillas the size of Nebraska). Home bakers use a rolling pin to flatten the dough into a pancake on a heavily floured board. Unlike corn tortillas, these wheat flour cakes can be made into various sizes. You can get an even consistency without cracks, whether the pre-flattened ball is the size of a golf ball or a tennis ball.

Why is the consistency of the flour dough more flexible and resilient than corn? The answer is gluten, a protein found in wheat doughs (and in some others, such as barley, rye, and oat). The kneading process used in tortilla preparation enhances and strengthens the formation of gluten, and Rolando Flores, research-food engineer at the United States Department of Agriculture, explains why gluten is so important:

> The major difference between the dough for corn tortillas and the dough for wheat tortillas is the difference between the corn and wheat proteins. The wheat gluten is the protein that gives the flexibility necessary to the tortilla dough so that it can be extended by manual or mechanical means. In some cases, bakers add up to 15 percent wheat flour to the corn tortilla.

Presumably, the wheat flour is added to strengthen the integrity of the corn tortilla dough, to avoid dreaded taco spillage.

So we have arrived at the inescapable conclusion that flour tortillas are larger because they can be made larger! In Mexico, the most popular size for flour tortillas is six inches, the same size as corn tortillas. But in the United States, restaurants always think about supersizing. Why offer a burrito the size of a taco when you can serve a gigante burrito, a quesadilla on steroids, or a taco salad served in a fried-flour tortilla "bowl" the size of a soup tureen? Surprisingly, flour tortillas outsell corn tortillas in Mexico and corn bests flour in the United States, according to Samuel Rodriguez, national sales manager of Olé Mexican Foods. In Mexico, the six-inch-diameter flour tortilla dominates the market, but even in the United States, the small size bests the large ones in sales, even if this doesn't jibe with our experience in restaurants.

The solution to this sub-Imponderable is that the Mexican market within the United States tends to buy the small-sized flour tortillas for home use, while Anglos tend to buy the large sizes. Tortillas are literally the biggest thing since sliced bread, since only sliced bread outsells tortillas in the United States. The Tortilla Industry Association proudly boasts that tortilla sales now exceed those of bagels and muffins, and tortillas now constitute a $6 billion industry in the United States alone, which does not even include the consumption of tortilla chips.

Submitted by Roger G. Reese of Los Alamitos, California. Thanks also to Douglas Watkins Jr. of Hayward, California.

Why Do Our Noses Run in Cold Weather?

In *How Do Astronauts Scratch an Itch?*, we discussed why kids get more runny noses than adults. The explanation lay mostly in the greater propensity of children to catch colds and infections. But sharp-eyed *Imponderables* readers noted that our noses run during cold winters even when we are feeling terrific, and wondered why even graybeards dab their noses in the winter.

Otolaryngologist Dr. Steven C. Marks, on behalf of the American Rhinologic Society, explains the physiology:

> The nose and sinuses are lined by a mucous membrane that contains both mucus-secreting glands and small cells called goblet cells, which also secrete a component of mucus. This mucus is produced in normal mucous membranes and in those that are infected or inflamed.

Many medical problems, such as viral or bacterial infections or allergies can cause your nose to run. But the nose's response to cold is a little different, as Marks explains:

The nasal and sinus mucous membranes are innervated [stimulated] by nerves which control, to some extent, the rate of mucus secretion. The response of the nose to cold air is in part a reflex mediated by these nerves. The cold air is sensed by the mucosa [mucus membrane], which then sends a signal back to the brain, which then sends a signal back to the mucosa: the result is a secretion of mucus.

What good does a runny nose do anyone but Kleenex? Keith Holmes, an ear, nose, and throat specialist from Dubois, Wyoming, believes that it is "a natural physiologic phenomenon of the organ to protect the warm lining of the nose," as cold irritates the mucous membrane. Marks speculates that "the increased mucus flow may be necessary to improve the humidification and cleaning of the air in the cold environment."

Submitted by John Miller of Lacona, New York. Thanks also to L. Gualtierie of Brampton, Ontario.

Why Do Bulls in Cartoons Have Nose Rings? And Why Don't Cows Have Them?

You give cartoonists far too much credit for imagination, David and Valerie. Long before there were punk rockers, bulls sported genuine brass rings.

The expression "bull-headed" wasn't pulled out of thin air. Bulls are among the most stubborn and least accommodating of farm animals. When a human wants a bull to move and a bull wants to sit for a spell, verbal commands are unlikely to work. Neither will a friendly little shove on the rear. Bulls' hooves have been known to crush the feet of owners who made them "see red," and they love to kick, too. Those rings are inserted to allow owners to "lead them around by the nose." As Richard Landesman, a University of Vermont zoologist, puts it, "Any tension on the ring will produce pain, and this can be used as a means either to train or restrain the bull."

Most bulls are "ringed" before they are a year old, in a procedure that isn't as delicate as a human ear piercing. For some reason, bulls don't welcome a veterinarian driving a steel rod through their septum, so they are given a local anesthetic and placed in a "head bail" to keep them from moving at an inopportune time.

Nose rings come in various sizes, and it is not uncommon for bulls to graduate to a larger-sized ring (as big as three inches in diameter) as they grow. Dan Kniffen, of the National Cattlemen's Association, told *Imponderables* that there are even temporary clip-on rings, called "bugs," the bovine equivalent of clip-on ties, that don't pierce the nostrils. Bugs can be used to shepherd the recalcitrant bull that needs to be moved occasionally.

It behooves the prudent farmer not to make permanent enemies of five- or six-ton creatures. Once the ring is inserted, vets urge owners to restrain the bull with a halter around his head, with a lead rope fed through the ring. Yanking on the ring directly is quite painful, and bulls have been known to carry a grudge. Hudson, New York, veterinarian Andrew S. Ritter also warns of the dangers of tethering a bull by the ring, lest it rip through the cartilage of the nose attempting to get free.

Why don't cows have nose rings? Although the occasional delinquent cow (and horse) sports a nose ring, the distaff bovines generally have a sunnier, more docile disposition than their bull-headed mates.

Submitted by David Ng of Capiague, New York. Thanks also to Valerie Valenzo of Chicago, Illinois.

Why Is Salt Sold in Round Containers?

If the more specific question, "Why is Morton Salt sold in round containers?" is answered, all will be revealed, for Morton has always dominated the sales of household table salt. Until the twentieth century, Morton sold its salt in cloth bags. Moisture infiltrated the bags with ease, leading to hard lumps. Consumers had to break up the caked salt (pounding the bag on countertops and pummeling the lumps with mallets were two preferred methods), and then put the salt into their own glass jars.

Morton experimented with square cartons at the turn of the century, which were more durable than cloth, but still didn't solve the caking problem. Plus, salt stuck to the corners of the square box. The first round carton was introduced in 1900, but it had its own nuisance—a wooden spout that had to be plugged in between uses. Not until 1911 did Morton's research discover the wonders of adding a smidgen of magnesium carbonate, which absorbed moisture and allowed the free flow of salt even in humid conditions (Morton now uses calcium silicate for the same purpose).

The only downside to the spiffy round container was that it cost more to manufacture than square ones, and the expense had to be passed on to consumers. The solution? Advertising.

The Morton Salt Girl, complete with umbrella, was introduced in a 1914 *Good Housekeeping* ad trumpeting the company's primary product benefit ("When it rains it pours"), and consumers proved willing to spend more money for free-flowing salt in the spiffy, blue round containers that Morton calls "cans" internally. Morton has held on to a dominant market share, the slogan, and a similar-looking container ever since. The shape has become so identified with table salt that other brands, including generic and store-brand competitors, have copied the packaging.

But there's no inherent advantage to the round shape. In fact, Morton representative Don Monroe told *Imponderables* that the company still sells bales and bags of salt for institutional uses. For example, a pickler might buy a twenty-five-pound bale of salt and dump the whole bag into a vat to brine cucumbers. Morton and other companies sell specialty products, such as kosher, pickling, and tenderizing salt in rectangular packages.

Submitted by Venia Stanley of Albuquerque, New Mexico. Thanks also to Ronald C. Semone of Washington D.C.; and Kathy Farrier of Eugene, Oregon.

Why Is There a Two-Minute Warning in American Football?

We almost didn't research this Imponderable because we assumed that the two-minute warning was instituted at the behest of the television networks, who wanted to make sure there were plenty of opportunities to plaster a block of commercials at critical points in the game—right before the climax of the first half and the end of the game. But we were wrong.

We regret ever thinking that the fine executives of professional football and broadcasting might ever be motivated by anything as crass as the mighty dollar. The two-minute warning debuted in 1942, and was created to remedy a nagging problem that threatened the fairness of the game. Until 1942, the official time was kept on the field, and scoreboard clocks often bore little resemblance to the official time. According to Faleem Choudhry, a researcher at the Pro Football Hall of Fame in Canton, Ohio, before the two-minute warning, scorekeepers had to notify each team when there was somewhere between ten and two minutes left in the game.

The looseness of the rules constrained coaches. Bob Carroll, executive director of the Pro Football Researchers Association, e-mailed us about the implications:

Obviously, it was important for a team in the closing minutes to know exactly how much time was left so it could make critical substitutions, stall, try to run out the remaining time, etc. Although the players on the field could ask the official, it took time to notify the bench.

On the other hand, taking time after each play to go over to each coach would have required stopping the clock after each play—possibly to the detriment of one team. I think the two-minute warning was a compromise that allowed the coaches to know exactly how much time was left and then keep a relatively accurate record on the bench.

These days, teams spend a part of most practices running their "hurry-up" offenses (sometimes known as a "two-minute offense"), a prearranged sequence of plays that require no huddle and are designed to burn off as little time as possible. Often the hurry-up offense will commence with the first play following the two-minute warning—after the more than two minutes of TV commercials, of course.

Submitted by Jim Welke of Streamwood, Illinois.

Why Does the Japanese Flag Sometimes Have Red Beams Radiating from the Sun?

Americans pledge allegiance to the flag. We salute the flag. We burn the flag. We try to pass constitutional amendments to criminalize burning the flag. We fight for the flag. We die for the flag.

A flag is an icon, imbued with emotions, dreams, and fears that extend far beyond its cloth and dyes. You can tell quite a bit about a country by its attitudes toward its flag. Case in point: Japan.

You are probably all familiar with the Japanese flag, the *Hinomaru* ("sun disc"), which some trace back to the time of the Emperor Monbu in the early eighth century.

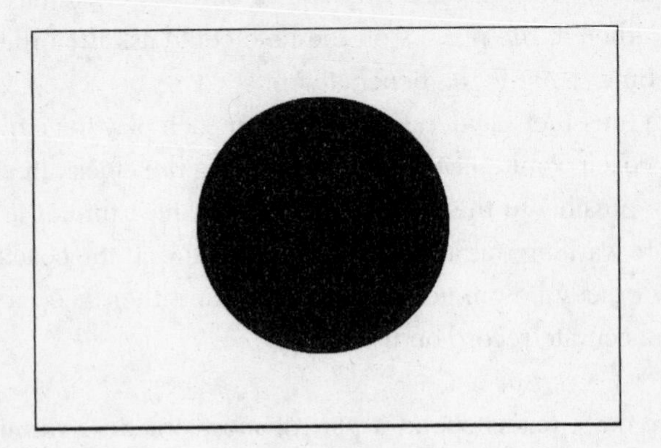

According to Dan Scheeler, librarian at the Sasakawa Peace Foundation USA Library in Washington D.C.:

> Legend has it that a priest named Nichiren presented a sun flag to the shogun at the time of the Mongol Invasions [launched by Kublai Khan] in the late thirteenth century.

During the fifteenth and sixteenth centuries, when various clans and military figures were vying for control of Japan, *Hinomaru* were displayed as military insignia (sometimes with different color schemes to designate particular factions, but the red sun on white background was most common).

For almost 300 years, Japan isolated itself from the West, but in 1853, the militias of two feudal lords fought and killed some sailors from the Royal Navy of England. One of the clans, the Satsumas, fought under the *Hinomaru,* and the English mistakenly assumed that this was their national flag. In order to avoid being mistaken for foreign vessels, a shogun agreed that it might be advisable for Japanese vessels to all carry the same flag, as Richard Allen Jones, of the Japan Information Center of the Consulate General of Japan, explains:

The flag, in its present form, was suggested by Lord Narakira Shimazu, head of the powerful Satsuma clan in southern Japan. The first display of the sun flag as the symbol of the nation was on the occasion of the trip to the United States, in 1860, of the first diplomatic delegation ever sent abroad by the Japanese government. The *Powhatan*, a United States Navy cruiser, was placed at the disposal of the Shogunate for this purpose. The ship flew the American flag at the stern and the Japanese flag at the bow.

None of this activity likely had the slightest effect on the average Japanese person. But things were about to change. In 1868, the Tokugawa Shogunate was stripped of power, and the Emperor Meiji assumed power. In January 1870, the prime minister proclaimed that all ships must fly the *Hinomaru*, and mandated the dimensions of the flag, which still remain the same today. According to Richard Allen Jones, the first time the *Hinomaru* was flown at a national ceremony was in 1872, on the occasion of the opening by Emperor Meiji of Japan's first railway. For more than a hundred years after the beginning of the Mejii era, Japanese citizens might fly the flag on important holidays, yet the *Hinomaru* did not possess great iconic value, possibly because it wasn't officially the national flag of Japan.

Soon after Meiji's reign, a special flag for the Imperial Navy was introduced. Beginning in 1889, naval vessels flew a sun disc flag with sixteen rays extending to its borders:

Masahiko Noro, executive director of the Japan Foundation in New York, wrote *Imponderables*:

> This flag was used by the Japanese military, particularly the Japanese navy, from the Meiji era until the end of World War II. The Japanese have not used this flag to represent their country since 1945.

The naval flag, then, is not another version of a national flag. Indeed, there really was no official national flag to be a variation of! So this Imponderable was based on an understandable misconception. We mistakenly assume that the "ray" flag is a variation of the national flag, when it is not related. Most Americans are familiar with the "ray" flag because so many depictions of the Japanese flag we have seen come from war movies, specifically World War II movies, where the naval flag is (realistically) depicted as the military ensign of the Japanese warriors. Even more confusing, the Treaty of San Francisco, which settled the conclusion of World War II, mandated that Japan eliminate its armed forces, so there wasn't much need for the naval flag after 1945. But in 1952, Japan started to build up "self-defense" forces, which looked suspiciously like a navy to most foreigners. In 1954, the Japanese Maritime Self-Defense Forces reclaimed the naval ensign as its own, and it still flies today.

Ironically, while the military flag's reemergence stirred little passion in the West, the Japanese exhibited deep ambivalence about the *Hinomaru*. To leftists, intellectuals, and union members, in particular, the flag represented a period (1931–1945) of xenophobia and unjustified military aggression. Even those without such strong political feelings tended not to be preoccupied with displaying the flag—even government buildings often did not fly the sun disc.

It was not until 1999 that the *Hinomaru* was officially proclaimed the national flag of Japan, and even then, only a particularly sad incident prompted the change. In Hiroshima, a high school principal was unsuccessfully attempting to use the *Hinomaru* and a patriotic song, the "*Kimigayo,*" in the school's commencement ceremony. Teachers objected not just to the flag, but also the song, which they thought glorified the imperial system, which was responsible for abhorrent military practices during the war. The lyrics do give credence to the grievance:

> May the reign of the emperor continue for a thousand, nay, eight thousand generations and for the eternity that it takes for small pebbles to grow into a great rock and become covered with moss.

The principal, caught between a school board that wanted the flag to be flown and the song to be played at commencement and faculty that was balking, committed suicide the day before graduation. Within days, the government pushed for legalizing the *Hinomaru* as the official national flag (and the "*Kimigayo*" as the national anthem) of Japan, and within six months, accomplished the task.

The brouhaha over the resurrection of the sun disc flag proves that it had become an icon in Japanese culture. Even for those who disagreed with making the *Hinomaru* the official flag, the symbol became worth fighting for.

Submitted by Dr. J. S. Hubar of New Orleans, Louisiana.

Why Do We "See Stars" When We Bump Our Head?

Want to see stars? We heartily recommend going to the countryside, where there are few lights, and looking up, especially on a cloudless night. As Gerry Goffin and Carole King so eloquently phrased it in their song "Up on the Roof," "At night the stars put on a show for free."

If looking up in the sky isn't edgy enough for you, you can try conking your cranium. Not that it costs anything to bump your head, except potential medical bills. But a knock on the noggin isn't as reliable as the sky or a visit to the local observatory—you may or may not see stars. And then there's the little matter of the ensuing headache.

We consulted with several neuro-ophthalmologists, as we weren't sure whether we saw stars because of damage to the eye or damage to the brain. As it turns out, there is a bit of controversy on the subject, but most agree that most of the time, the "eyes have it."

Lenworth Johnson, a neuro-ophthalmologist at the Mason Eye Insitute at the University of Missouri, Columbia, wrote to us that when you bump your head, you shake the vitreous gel in the eye. The vitreous gel, also known as the vitreous humor, is a transparent, colorless jelly that fills the eyeball behind the lens (front part) of the eye. The vitreous, which is adherent to the underlying retina, then "jiggles the retina." The retina is the sensing element of the eye that sends information to the brain about light, color, and brightness. This jiggling sends the signal of stars to the brain. As Johnson analogizes it:

> This is equivalent to having your skin squeezed and reporting pain or other sensation because of the nerves in the skin sensing the touch.

The jostling of the retina doesn't translate to pain, though, as Scott Forman, associate professor of ophthalmology, neurology, and neurosurgery at New York Medical College, explains:

> Seeing stars after a head injury probably refers to what is known in the profession as "Moore's lightning streaks." A gentleman by that name coined the term to refer to the visual disturbances arising from sudden acceleration/deceleration of the eyeball (that accompanies a blow to the head). This produces a gravitational force exerted on the vitreous body, or vitreous humor.
>
> When the vitreous tugs on the retina, as it would if a sudden force is applied to it, the retina wrinkles ever so slightly. The mechanical deformation of the retina is not felt as pain, since there are no pain fibers in the retina. However, it sets off, most likely, a wave of depolarization, a change in the electrical charge or electrical activity of the photoreceptor layers, those layers of the retina containing the elements that receive light from our environment. The mechanical deformation has the same effect as light entering the retina. That is, it sends a signal to the optic nerve and hence the visual brain (the occipital lobe, eventually, and other visual "association areas" in the brain) that we interpret as spots of light, sparkles, lightning, or whatever.

If you want to see *someone else* see stars, your best shot is a Warner Brothers cartoon—many an occipital lobe has been banged by a hammer in cartoons. But athletes, accident victims, and crime sufferers who undergo concussions (the soft brain knocking against the hard surface of the skull) also often see stars. One study indicated that nearly 30 percent of athletes who suffered head injuries from direct impact saw stars or unusual colors. Epileptics sometimes see stars during seizures, and sometimes after them.

So you can see stars by disturbing the brain directly, and leaving out the "middle man" (the retina), but chances are that, unlike bunnies in a cartoon, any bop on the head that is strong enough to make you see stars is sufficient to send you to an emergency room.

Submitted by Stu Levy of Seattle, Washington.

How Is the Maximum Occupancy for a Public Room Determined?

In our first book, *Imponderables,* we pondered who in an elevator was responsible for determining the weight of fellow passengers. Just as elevators have posted weight limits, so do public rooms have signs indicating the maximum occupancy permitted by law. Somehow, the numbers have always seemed capricious. Why are nightclubs packed to the gills, while our book signings look like ghost towns? (We're assuming that throngs of folks are kept at bay at the bookstore door to avoid overcrowding and overzealous fans.)

At the most basic level, the determination of occupancy load is cut and dried. As architect Norman Cox, of New York's Franke, Gottsegen, Cox Architects told us:

> The most common method is to divide the total floor area by some specified quantity of square feet per person in order

to determine the maximum number of occupants. This quantity varies usually between four and twelve square feet per person, depending on seating type. For example, in a restaurant with tables and loose chairs, the occupancy of a 1,200-square-foot room might be 1,200 divided by 12, equaling 100 persons.

Maura Gatensby, a Vancouver, Canada, architect, shared some of the occupant loads for her area (numbers have been converted from square meters to square feet and rounded off to the nearest hundredth of a square foot):

standing space—4.30 square feet per person
grandstands—6.45 square feet per person
bowling alleys and pool halls—100.10 square feet per person
classrooms—19.91 square feet per person
dining rooms, cafeterias, and bars—12.91 square feet per person
shops—49.513 square feet per person
offices—100.10 square feet per person

The square footage allocated per person partly determines everything from the required numbers of exits and resources for handicapped persons to the number of washroom fixtures in the lavatories. Depending upon the locality, the police or fire department might enforce these rules.

There is no single standard for building codes—states might use BOCA (Building Officials and Code Administrators), UBC (Universal Building Code), or SBC (Standard Building Code). Big cities tend to have their own codes, and might use one of these three codes as a model and modify it.

Fire departments can exert pressure on buildings, too. In some localities, if a fire department finds that the sprinkler system in a building isn't adequate, or that the stairways are not wide enough to handle emergency situations, the occupancy load of a building can be reduced from the norm.

The insurance industry is breathing down the neck of contractors and facility owners, as well. Even if the city building officials and fire departments sign off on an occupancy load, an insurance company might not be willing to underwrite insurance unless stricter guidelines are enforced. In many businesses, more bodies in a room mean more moolah, so the push and pull of commerce with safety and insurance is a never-ending battle.

Submitted by Stephanie Englin of Tukwila, Washington.

Why Do Teddy Bears Frown?

Considering that the toy business is full of sugarcoated images for children, and the happy face is the default countenance for dolls and most stuffed animals, we've often wondered why teddy bears are downright dour. So we contacted teddy bear artists, designers and manufacturers, hardcore collectors, and folks who write about teddy bears for a living to illuminate exactly what is bumming out stuffed bears.

Strangely, the first teddy bears were made in Germany and the United States in the same year—1902. Mindy Kinsey, editor of *Teddy Bear and Friends* magazine, picks up the story. At the beginning, at least, teddy bears were designed to appear realistic:

> In Germany, they were modeled after bears Richard Steiff saw in the zoo and at the circus. In America, teddies were inspired by the bears Theodore Roosevelt hunted (and in a particularly famous instance, failed to shoot) and were named after the president himself.
>
> Early teddies, therefore, had long muzzles, long arms,

humped backs, and small ears, much like the real thing. Their mouths tended to be straight embroidered lines that might appear to frown, but were only meant to mimic their real-life counterparts.

When we called the big teddy bear makers, such as Gund, Steiff, and Russ Berrie, the designers couldn't articulate why the expression of most of their bears was sad. Some suggested that they weren't *trying* to make their bears frown at all. But go to the Web site of these companies, or visit your local toy store, and we think you'll agree that compared to most other stuffed animals and toys, the classic bears could use a dose or two of Prozac.

But not just designers denied the "frown" premise. Kinsey's response was typical:

> Today's teddy bears, however, can have big grins, wistful smiles, laughing open mouths, puckers, and every other mouth imaginable. Some still have the straight-line mouths, but I like to think of them as wise, contemplative, trustworthy, or sincere expressions—not frowns.

Jo Rothery, editor of the English magazine *Teddy Bear Times*, thinks variety is the spice of teddy bear life:

> Of course there are some bears that are definitely grumpy and have been designed that way by the bear artist. Some collectors do specialize in the grumpy characters, perhaps because they remind them of someone—fathers, husbands, grandfathers, colleagues, etc. And other collectors, particularly of vintage teds, feel that a sad expression adds to the character of the bear and reflects his age and all the experience he has had over the years.
>
> There are some very "smiley" bears, whose mouths are upturned and instinctively make you want to smile back when

you look at them. Again, there are collectors who specialize in such bears, but I think the majority of us like to have a collection that includes lots of different expressions, possibly even some of the openmouthed variety, although it is hard to get that particular expression right. Some bear artists succeed in capturing that "wild" natural look very well without making the bears look at all scary.

Rothery adds that it is just like when one dog in a litter stirs your heart, even though "they all look alike."

> Even when you see a lineup of identical teddies, each one will have a slightly different expression, and there is one that will appeal to you more than any of the others and demand that you take him home with you.

We were shocked when three Gund designers couldn't articulate why they drew bears' expressions the way they did, but one creator, Linda McCall, of Key West, Florida, describes it as an almost mystical process:

> Some of my bears' mouths smile, some frown, and some look really, really grumpy! It just depends on the "feel" I get from the bear. I know it sounds strange to someone who probably has never made a bear. You stitch the darn thing together and you let it sit overnight. Then you look at it again and you just know if it should be a happy bear, a thoughtful bear, or whatever mood it seems to convey. That's why if you look at all artist-made bears, no two would ever be alike.

McCall, and several other sources, think the tradition of the "frowning bear" stems from an attempt to mimic how a real bear's mouth looks. After all, bears in the wild aren't known for their grins. The "realistic" theory, perhaps the favored one among our sources, contends that most bears aren't frowning, but merely exhibiting a neutral emotion.

If you look carefully at the faces of teddy bears, you'll see that the mouths of many are shaped like an upside-down capital Y. Teddy bear artist Cherri Creamer, of Alive Again Bears, says that the inverted Y is used to align the face so that the nose and eyes conform to the mouth. Whatever reason, the inverted Y provides a downward cast to bears' mouths. So this "convenient" method of aligning the bear might be responsible for what we interpret as a frown.

We're partial to a psychological theory to explain the "frown" of the teddy bear. Jo Rothery comments that the inverted Y

> gives teddy bears a contemplative, relaxed look, an expression that makes them seem only too willing to sit there, "listen," and absorb their owners' emotions, whether those emotions happen to be sad or happy.

If the emotions of a teddy bear are opaque, a child can pour his emotions into his plush toy, and the bear becomes an instant empathizer. Marc Weinberg and Victoria Fraser, creators of Ballsy Bear and Bitchy Bear, agree and comment:

> Teddy bears have been adopted for the most part by children. They're the child's only security blanket. You seek him out when you need comfort, when you need a shoulder to cry on. Teddy bears reflect the feelings of their owners.

If bears reflect the feelings of their owners, then Ballsy Bear's customers are not a happy bunch. Their two bears, Ballsy and Bitchy, are none too happy campers, and their classic inverted-Y mouths reveal not just a frown but a scowl. Created by a husband-wife team who both were victims of Internet startup failures, Ballsy and Bitchy are not cuddly and are happy to let you know:

Who needs another teddy bear that says, "I wuv you." Give your friends, family, and loved ones the gift that says what you really mean: Ballsy Bear and Bitchy Bear—the World's Nastiest Talking Teddy Bears!

Submitted by Tim Walsh of Ramsey, New Jersey.

Why Are the Sprinkles Put on Ice Cream and Doughnuts Also Called Jimmies?

Do you call Coca-Cola "soda" or "pop"? Do you call those overstuffed sandwiches "hoagies" or "submarines" or "torpedoes" or "grinders"? The answer depends upon where you're from. Although we've received this "sprinkles" versus "jimmies" Imponderable before, we weren't too excited about researching another question about regional differences in food names until we received this e-mail from reader Netanel Ganin:

> I work at an ice cream store where there is currently a hot debate that has spread to most of my friends. Is the term "jimmies" for those sprinkles that people put on ice cream a racist term? What are its origins?

We had never considered "jimmy" to have racist connotations, so we decided to do some research. Dictionaries weren't of any help. The *American Heritage Dictionary*'s definition was typical: "Small particles of chocolate or flavored candy sprinkled on ice cream as a topping. Etymology: origin unknown."

But quickly we could pinpoint the epicenter of jimmydom to the Mid-Atlantic and lower northeast United States. While ice cream parlors in San Francisco or Atlanta offer "sprinkles," in Boston or Providence, you are likely to be proffered "jimmies." You can buy jars of the cylindrical candy today, but most manufacturers hedge their bets, describing the product as "sprinkles/jimmies." Clearly, the terms are used interchangeably, although in some localities, "sprinkles" is used to describe all flavors but chocolate, and "jimmies" for chocolate sprinkles.

There might be confusion about what to call the darn things, but we can trace their history to one man—Samuel Born. A Russian immigrant, Born settled in San Francisco and within a few years made his first contribution to culinary culture—the Born Sucker Machine, which mechanically inserted sticks into hard candies—lollipops entered the twentieth century.

In 1917, Born opened a candy store in Brooklyn, New York. He trumpeted the freshness of his confections by using the slogan "just born." Along the way, Born invented the chocolate coating for ice cream (the kind used to enrobe soft ice cream at Dairy Queen, Carvel, Foster's Freeze, etc.). But for our purposes, the key invention of Sam Born was the chocolate jimmy.

Nearly a century later, the grandchild of Sam Born and his cousin are co-presidents of the company, now called Just Born. Janet Ward, of Just Born, proudly proclaims:

> Yes, jimmies were invented at Just Born and we have in our archives some of the advertisements from that time period and containers with the word "jimmies" and the Just Born logo on them. Although there is nothing in writing to confirm it, it is commonly known here that the chocolate sprinkles were named after the Just Born employee who made them.

According to some sources, that employee was Jimmy Bartholomew, who went to work at Just Born in 1930 and labored at the machine that produced the chocolate pellets that have blanketed many an ice cream and doughnut ever since, but the current co-president of the company, Ross Born, can't positively confirm "Jimmy's" last name. Back then, most ice cream parlors offered jimmies for free, so they proved most popular with customers who wanted an extra sugar rush without springing for the cost of a sundae.

Just Born has gone on to sell many well-known products, including Marshmallow Peeps, Mike and Ike, Hot Tamales, and Goldenberg's Peanut Chews. But sadly, Ross Born told us: "We stopped producing jimmies in the late 1960s; it wasn't one of our leading items."

The jimmies produced by Just Born were always brown, but not necessarily chocolate—perhaps "chocolate-looking" would be more accurate. We've noticed that most of the jimmies we've tasted have a waxy, gummy consistency and an off-taste that doesn't resemble chocolate. Ross Born reveals a little-known secret about jimmies—they were probably never made from real chocolate:

> The forerunner of jimmies was a product called "chocolate grains," which was a chocolate product. As I understand it, jimmies were non-chocolate, at least the formula that I recall and is stated on the container we have in our archives. However, jimmies look like chocolate, and most people would call it that.

Assuming that jimmies were named after Mr. Bartholomew, we can assume he was of Irish descent. So then why is there a fear that "jimmy" is a slur against African Americans? Some etymologists speculate that "jimmy" is a variant of Jim Crow, the title character in a famous minstrel song performed by black performers (and white performers in blackface) in the 1830s. The song became so popular that anything that pertained to African Americans was dubbed "Jim

Crow," especially in racist contexts. Later, "Jim Crow" came to refer to segregation of blacks from whites, including the infamous Jim Crow Laws, which were enacted in the South to preserve segregation after the Civil War.

But this theory seems lame to us. Although "jimmy" has many slang connotations in American English, from a crowbar to an engine made by General Motors, none of them refer specifically to blacks. We subscribe to a much simpler explanation. In most, although admittedly not all, places where sprinkles are called "jimmies," the reference is only to chocolate candies. Since their color resembles the complexion of many African Americans, it's easy to see how jimmies might have picked up the racist connotations, even if the inventor of jimmies intended only to honor his Irish American employee.

Submitted by Kendra Delisio of Chelsea, Massachusetts. Thanks also to Rick Kot of New York, New York; and Netanel Ganin of Sharon, Massachusetts.

Why Do We Draw a *Bead* on a Target?

It doesn't sound too intimidating to sketch a little picture of a bauble on a target, but that isn't the "bead" in question. The bead referred to in this phrase, coined in the mid-nineteenth century, is the foresight of a rifle, which, come to think of it, looks like a tiny bead.

Submitted by James Gleick of Garrison, New York.

Why Do Many Dictionaries Say That the Days of the Week Are Pronounced "Fri-DEE," "Sun-DEE," etc.?

In May of 1992, we received this plea from reader Richard Jackoway, of Shell Beach, California:

> I realize this does not sound like a tough question, but go to almost any dictionary and I'll bet it says the way you have said the days of the week all of your life is wrong.
>
> I have before me the *American Heritage College Edition* (picked simply because it was the closest), and if I look up Monday it says, "Monday (mun'dē, 'dā). I have been doing this for years and almost every dictionary shows Monday (or any other day of the week) with this unusual preferred pronunciation for the "day" syllable.
>
> I don't know about you, but I've never known anyone to pronounce the days of the week like this. Are all of my friends and acquaintances out of step or is this some obscure practical joke by lexicographers?
>
> As noted, there are a few dictionaries that have the "normal" pronunciation first, but their numbers are few. Can you solve this linguistic Imponderable?

In 1992, we were dumbfounded, assuming that the premise of this question must be wrong. We consulted the two dictionaries closest to us, and sure enough, both our *Merriam-Webster* and *Webster's New World Dictionary* listed "dē" before "dā." Always eager to help the desperate, we did look into this Imponderable, and reached the estimable Brian Sietsema, who was then the pronunciation editor of Merriam-Webster dictionaries. Sietsema, who had the good humor of one who has heard similar rantings from enraged readers, held his ground. "Most people actually do say "dē," he insisted, much to our amazement.

"How do you decide what pronunciations you list?" we asked, wondering what role liquor or hallucinogens played in the mix. Sietsema proceeded to indicate that orthoepy (the study of pronunciation) is far from a hard science. Until the 1930s, a pronunciation editor's job was to act as judge and jury of how words should be pronounced; but with the advent of radio as a mass medium, Merriam-Webster decided that its role should be to describe how people are really speaking, rather than how an editor, or speech professors thought they should—Sietsema proudly followed in this new practice.

Part of Brian's day consisted of listening to the radio, watching television, and writing down phonetic citations as they arose, keeping a running file of variants. Sietsema emphasized that today's mispronunciation could be tomorrow's correct pronunciation, and he gave some examples. Decades ago, the word *iron* was pronounced "I-ron" instead of "I-ern." The toothpaste additive "FLU-er-ide" morphed into "FLOOR-ide." And in a variant listed in *Merriam-Webster* that caused a firestorm, "NU-kyuh-ler" was listed as an alternative pronunciation, even if it wasn't deemed as an "acceptable" one.

Somehow, we got sidetracked and tackled more pressing Imponderables, like "Why Did Pirates Wear Earrings?" and never finished our research. Now, more than ten years later, we posed our Imponderable to Constance Baboukis, managing editor of the U.S. Dictionaries Program of the Oxford University Press, and she replied:

> This is an easy question and no mystery at all. The -DEE pronunciation for the days of the week (but not other -day words like holiday) used to be the only pronunciation, and is probably still used by more people in the U.S. than the -DAA pronunciation.

Agggh! We went to the Merriam-Webster Web site and listened to the disembodied voice pronounce "Fri'DEE," (www.m-w.com/cgi-bin/dictionary?book=Dictionary&va=friday) and what came back sounded like nothing we've ever heard.

But dictionary editors ain't no dummies. We figured there must be more to this issue. So we gathered an all-star team of "pronsters" (the favorite nickname for orthoepists), all of whom have advised on or edited many dictionaries:

Constance Baboukis, (CB) of Oxford University Press
Bill Kretzschmar, (BK) professor of English and linguistics at the University of Georgia and editor of the *Oxford Dictionary of Pronunciation for Current English*
Rima McKinzey, (RM) freelance orthoepist, co-editor of the mammoth *Pronouncing Dictionary of Proper Names*
Enid Pearsons, (EP) a lexicographer with a specialty in pronunciation and former senior editor in charge of pronunciation for Random House dictionaries.

Of course, we found that these orthoepists were the opposite of dummies, and as editors were faced with constraints so daunting that we could almost forgive them for "Mon-DEE." Dictionary editors, for the most part, gather pronunciation data just as Sietsema did—by the seat of their pants:

RM: Those of us in the field just try to listen a lot.
BK: Most dictionary pronunciation editors just rely on their own experience, and this is not too reliable. I have seen some editors ask for comments on e-mail lists, and so some of them are getting anecdotal reports from people who write back to the list.

One fear is that because most U.S. dictionaries are based on the East Coast, and editorial staffs tend to be populated by well-educated Easterners, this "bias" might slip into the pronunciation guides. Kretzschmar is the director of the American Linguistic Atlas Project, which has been gathering data on how "real people" speak in different regions of the country for nearly seventy-five years.

Pronunciation editors don't have tin ears. Their excellent ears, ironically, are part of the reasons why we see "DEE" as an acceptable pronunciation. All of the dictionaries we consulted list "dā" as the proper pronunciation for "day," but:

RM: Just because an entry consists of one or more free forms (i.e., "day," in the names of the days of the week) it doesn't necessarily follow that they will be pronounced as they are when they're alone. Cupboard is not pronounced "KUP-board," for instance.

BK: In "Monday," it's a matter of stress: primary stress is on the first syllable; the second syllable can either have secondary stress, in which case it still has the -dā pronunciation, or it can be relatively unstressed, in which case the vowel quality changes to something like -dē. Think of words like *tomato* or *potato*, where the last syllable sounds like O if stressed, or changes to something like UH if not.

EP: The final syllable in all seven days of the week lacks stress, and earlier American dictionaries recorded something like "MUN-dih, T(Y)ooz-dih," etc. to indicate that. That was back in the days when American dictionaries showed "HAP-ih" and "PRIT-ih" for *happy* and *pretty* as well. I remember changing that "ih" sound in the Random House dictionaries in the 1960s, on the theory that it more closely reflected a British pronunciation than an American one.

In short, there are way too many variants for any dictionary, even the unabridged ones, to deal with such nuances in the tiny amount of space (usually less than one line) in which pronunciation is relegated. Just consider all the other questions that pronunciation editors have to take into account:

1. *How Can Regional Differences Be Accounted For?*
Although Baboukis contends that a plurality of Americans

say "DEE," she notes that this isn't true in the East, "so it sounds strange to our ears." Likewise, Pearsons professed to have heard "DEE" spoken "innumerable times," but it is clearly more common in the Midwest (think of the protagonist of the movie *Fargo*), and "perhaps among older speakers."

The Northeast may make fun of southern or midwestern patterns of speech, but as McKinzey puts it: "Proper Bostonian speech is no more correct than proper Atlantan speech." England had a tradition of "received pronunciation," the style of the upper class learned in public schools (the equivalent of American private prep schools) and promoted at Oxford and Cambridge Universities. We have no such tradition, and there is little reason for dictionaries to exclude variants because they are the choices of a particular region.

2. *How Can Dictionary Readers Decipher Pronunciation Guides When There Are No Standard Recognized Symbols*?

RM: American dictionaries are all self-pronouncing. That is, however you pronounce the symbol the dictionary may be using to show the pronunciation of the *a* in *cat*, that is how you'll pronounce those entries. This is inherently different from most dictionaries around the world, which use the International Phonetic Alphabet (IPA). The IPA has one sound to one symbol. This works for most languages around the world, but not American English, for multiple reasons. We're not taught it in school, for one thing."

McKinzey points to a horrendous problem for pronunciation editors. Even educated American readers are likely to know the symbol for a short and long vowel, and little else about pronunciation symbols. Every dictionary's scheme is different, and we daresay the average dictionary user never consults the pronunciation guide at the beginning of the dictionary to sort out the problem.

Kretzschmar is a proponent of IPA, and works with it in the *Oxford English Dictionary* and his *Oxford Dictionary of Pronunciation for Current English*:

> **BK:** While *OED* and *ODPCE* render pronunciations *phonetically*, American dictionaries usually render them *phonemically*. . . . Americans usually pronounce *latter* and *ladder* the same, which *OED* and *ODPCE* represent, but most American dictionaries represent one with a *t* and one with a *d*. . . .
>
> We chose IPA for our *ODPCE* because it is clearly the best for the world market. Tradition is really the only reason for American dictionaries to retain their (exotic and inconsistent) symbol sets—the marketing departments of dictionary houses are afraid that people won't buy their dictionaries if they are the first to switch to IPA.

Of course, even if IPA is more accurate and consistent, marketing departments (and editors) have a reason to fear readers recoiling from the prospect of having to learn a new set of "strange symbols," only some of which look like English letters.

3. *Should Dictionary Entries Be Descriptive or Prescriptive?*

Is the purpose of a dictionary to record how a user should pronounce a word, or to record how ordinary people *do* pronounce a word?

> **RM:** Most dictionaries do try to be descriptive, both in definition and pronunciation.
>
> **BK:** Dictionaries should be witnesses, not judge or even jury (*pace American Heritage Dictionary*). Pronunciation is *much* more variable than we give it credit for. There are different habits of pronunciation that occur to varying degrees

within regional and social groups of people in America, and even particular individuals vary their pronunciation from use to use of the same word. Of course, this means that there is always a choice of what to include—some judgment has to be involved because no dictionary has room for so many variants.

Kretzschmar sees his role not unlike a cultural anthropologist—to record what exists in the real world. The *American Heritage Dictionary* is more prescriptive—and has amassed a "usage panel" with luminaries from Supreme Court justice Antonin Scalia to writer Susan Sontag to weigh in along with lexicographers on their opinions. Most dictionaries fall somewhere in between the Merriam-Webster dictionaries, which tend to be the most descriptive, and the AHD. Enid Pearson's position strikes a balance that appeals to us:

EP: As for prescriptivism vs. descriptivism, each dictionary has its own philosophy, and even that may not be internally consistent when it comes to language change in pronunciation, grammar, and usage. Mine is essentially descriptive (the language does evolve), but with the strong caveat that—as a lexicographer—when you're describing, you owe it to your dictionary users to make clear what attitudes still exist in the real world, what the older "standards" are among educated speakers, and what negative responses the users may encounter if, for example, they persist in saying "NOO-kyuhler," or they say, "Give it to Mary and I," or they spell *memento* with an initial "mo-."

I guess that makes me a cranky prescriptive descriptivist.

Kind of like a God-fearing atheist!

4. *If Variant Pronunciations Are Listed, What Order Should They Be In?*

Many readers assume that the definitions in a dictionary are placed in order of their frequency of use. But the first pronunciation listed in a dictionary is not necessarily the "correct" one:

RM: [Most dictionaries] try to place what the pronunciation editor considers the most common variant first, followed by less common or more regional variants. Many times, variants are equally common. However, a dictionary is a two-dimensional medium and one pronunciation cannot be placed on top of another and still have the entry readable and usable. Beyond that, editorial decisions are what makes "Eether" come before or after "EYEther."

BK: Order of pronunciation entries has different significance in different dictionaries. In the *Oxford Dictionary of Pronunciation for Current English,* we say explicitly that order of presentation should not be taken as indicating preference or frequency of the pronunciations offered. . . . But people often do interpret the order of pronunciations that way. You really have to consult the frontmatter of each dictionary in order to determine what, if anything, is implied by the order in that dictionary.

5. *When Pronunciations Are in Flux, When Should Dictionaries Change Their Entries?*

RM: Pronunciations change quite slowly and dictionaries are fairly conservative in their approach.

Sometimes, as is lately the case, a growing pronunciation shift is essentially ignored for the clarity of other pronunciation distinctions. The growing merging of the pronunciation of *cot* and *caught* or *Don* and *Dawn* will no doubt be ignored in American dictionaries because (1) where on the vowel spectrum they merge is still shifting and (2) whatever symbol is chosen will have a greater impact on other words not yet merged.

This is precisely what BK alluded to when he said that most American dictionaries render pronunciation phonemically, so that distinctions are not made between similar sounds, even if native speakers could distinguish them if pressed. Phonetic transcriptions attempt to describe every single sound made in a language—certainly not possible with the limited number of symbols with which dictionary editors must work.

But even if dictionaries are slow to amend, eventually they do, which is one reason we're not unhappy that it has taken us twelve years to finally answer this Imponderable. For while twelve years ago, it was possible to find dictionaries that listed only "Sun-DEE," or "Fri-DEE" before "Fri-Day," the times they are a-changin'.

CB: Current dictionaries show both, sometimes with the DAA pronunciation first because it is gaining ground.

Keep in mind that all dictionaries with new covers do not necessarily contain new content. Those bargain ones on the bookstore tables are often out-of-print oldies that were bought up cheap and refurbished, but not rewritten. If you use recent editions of American dictionaries written by active staffs who do research, such as Oxford, Merriam, New World, and American Heritage, you will get current pronunciations.

EP: Happily, most current American desk dictionaries show both pronunciations now—some with -DEE first, some with -DAA first, some with -DAA only. The -DAA pronunciation is, I suspect, taking over. Perhaps it is time your reader bought a new dictionary.

We recently contacted that reader, Richard Jackoway, and found out that he is a wordsmith himself—city editor of the San Luis Obispo *Tribune*. After being apprised of the *Reader's Digest* version of the explanation above, Jackoway reiterated his plaintive wail: Do the editors really think anywhere near as many people say "Mon-DEE" as "Mon-DAA"? Are they really suggesting that people pronounce the word "Mon-DEE"?

We looked at all the current major dictionaries and found that, as all of our orthoepists suggested, you absolutely must look at the pronunciation guides in the front of each dictionary to learn how each word should be pronounced. And if you want to know the significance of the order in which variants are listed.

We have paraphrased each dictionary's policy and rendered the pronunciation in our own scheme:

American Heritage Dictionary (does not indicate which variants are preferred or popular, merely that variants are included "whenever necessary"): "Fri-DEE, Fri-DAA"

Merriam-Webster Dictionary (no preference for variant order): "Fri-DEE, Fri-DAA"

Oxford American Dictionary (preferred pronunciation is listed first): "Fri-DAA"

American Century Dictionary (unless there is an explanation, either variant is acceptable): "Fri-DAA, Fri-DEE"

Oxford American Dictionary of Current English (the first listed is for "more frequent or preferred pronunciations"): "Fri-DAA, Fri-DEE"

Random House Webster's Dictionary (no stated policy about variants): "Fri-DAA, Fri-DEE"

Oxford English Dictionary ("The order of variants need not be one of decreasing frequency." Variants are shown "which can safely be regarded as allowable in British English at the present time, within the formal received pronunciation that does not give rise to any social judgement when heard by most native speakers."): using IPA equivalents of "Fri-DAA, Fri-DEE"

Penguin Webster's Handy College Dictionary: ("The criterion for pronunciations is the best usage in regions where there is no marked peculiarity of speech and in normal conversation rather than in formal speech." This is the only dictionary we've found that lists only one pronunciation per word): "Fri-DĖ" (the dotted *E* is defined as "sounding the *e* in *maybe* as opposed to the first *e* in *mete*).

Webster's New World Dictionary: ("Each variant pronunciation may be regarded as having wide currency in American English unless a qualifying note has been added to a particular variant indicating that it is less common.") "Fri-Da"

Victory! While the *American Heritage Dictionary* remains unmoved, *Webster's New World Dictionary* has not only pushed "Fri-DAA" to the front, but also banished "Fri-DEE" to "pron purgatory." In your lifetime, Richard Jackoway, perhaps *AHD* will see the light.

Submitted by Richard Jackoway of San Luis Obispo, California. Special thanks to Erin McKean, of Chicago, Illinois.

Why Do the Speed Controls on Fans Go from "Off" to "High" to "Low"? Wouldn't It Make More Sense for Them to Go from "Off" to "Low" to "High"?

Who would think that this humble mystery would be among the ten most-often asked here at *Imponderables Central*? Several readers compare fan controls to audio devices, which after all, don't go from "off" to "ten" to "one." The audio configuration saves a little energy and a lot of our residual hearing. When you are shutting off a piece of musical dreck, the last thing you want to hear is the noise at maximum volume right before you reach the exalted bliss of silence.

The analogy between radios or stereo system and electric fans (or air conditioners) isn't perfect, though. When you turn on a fan, you are usually uncomfortable. The room is too warm, or too stuffy, or too humid, and you want relief. As Don Thompson, an engineer at fan manufacturer Comair Rotron, put it: "If I turn on a fan, I want maximum cooling to relieve myself or perform a task. Immediately!"

If the maximum setting isn't strong enough to cool off the room, you need a stronger fan. If "max" is too much, that's what "low" is for.

Thompson calls this approach the "period of patience"—customers want maximum relief as soon as possible. When the zone is reached, the device is switched to a lower mode to decrease the noise and conserve energy.

The speed configuration isn't only for the benefit of us end users, though. Charles Richmond, vice president of engineering at cooling manufacturer EBM Industries, wrote *Imponderables* that the off-high-low configuration makes engineering sense. The greatest workload of a fan or air conditioner is right when it starts—when the motor must fight against inertia, the ambient air is the most stagnant, and the user's point of patience is leaning toward the impatient.

Think of a merry-go-round. Its motor faces its heaviest load when it starts to spin from a standing start; once it is turning at its normal operating speed, it requires much less work for the engine to maintain the same speed. If you started the engine at a lower power, you might not have enough juice to start the merry-go-round from a dead start. There may be no wooden horses or brass rings on a fan, but the principle is the same.

Submitted by Herman London of Fishkill, New York. Thanks also to Brett Holmquist of Burlington, Massachusetts; Josh Metzger of Hamilton, Ohio; Ned Smith of Menands, New York; Suzanne Amara of Boston, Massachusetts; Rob Shifter of Los Angeles, California; William Wimmer of Benton, Arkansas; Robert King of Grand Forks, North Dakota; Eric J. Roode of Claremont, New Hampshire; and John Chaneski of Hoboken, New Jersey; and many others.

Why Are Public Radio Stations Clustered on the Low End of the FM Dial in the United States?

As part of the Public Broadcasting Act of 1967, the United States Congress mandated that twenty channels on the new FM band be reserved for noncommercial, educational use—87.9 MHz through 91.9 MHz.

Of course, *we* know that men and women with pure hearts, untainted by avarice or ambition, populate our Congress. But more cynical folks might wonder why Congress would allocate such valuable "real estate" on the broadcast dial to educational, artsy-fartsy types who presumably had little lobbying power and even less money to contribute to political campaigns.

Maybe we should be more cynical. In reality, most of the pressure to lump the noncommercial broadcasters together on the FM dial came from commercial broadcasters who did have bucks to throw around. Prior to the 1967 allocation, many colleges ran "carrier current" AM stations, closed-circuit transmissions with low wattage that often managed to interfere with the signal of commercial stations. The big AM stations squawked about the problem, but the FCC did not have the manpower to police the problem.

But as Michael Starling, vice president of National Public Radio Engineering, explains, another issue was even more pressing:

> There were ongoing complaints about noncommercial stations that were assigned high-power AMs that were very desirable frequencies commercially. This was compounded by a growing number of mutually exclusive applications between commercial and noncommercial stations—something the commission had no way to resolve. This was an apples-and-oranges situation that pitted public-service, educational institutions against large commercial broadcast interests.
>
> Thus, the FCC thought this would be the ideal solution: set up a part of the new FM band to move the carrier current stations, which would clean up the carrier current interference without having to do battle with the Harvards, Columbias, etc., who had these low-power AMs. These became the Class-D 10-watt stations. It would also be the home of the future noncommercial stations, so that there would not be the previous

mess of trying to evaluate in a comparative context commercial and non-commercial interests.

The non-commercial stations are clumped at the bottom of the dial, where they can interfere only with one another's signals. Why were they put at the low end of the dial instead of the high? No one seems to know.

Submitted by Ed Katzmark of Superior, Wisconsin. Thanks also to Leonard Berg of Van Nuys, California; a caller on the Kerry Rodd Show, Minneapolis, Minnesota; and Tara Alexander, of parts unknown.

Why Do Pregnant Women Get Strange Food Cravings? And Why Do They Suddenly Start Hating Foods They Used to Love Before They Were Pregnant?

Is there something specific about being in the family way that produces a sudden passion for a pickle sundae? A pork-and-banana sandwich? Or Twinkies with a dollop of mustard?

A few social scientists subscribe to the notion that cravings are "all in pregnant women's heads," but the nutritionists and medical experts we consulted dissent. Food cravings are prevalent all over the world: We found scores of studies that found at least some food cravings in one-half to more than 90 percent of pregnant women, with most falling in the 65 to 75 percent range.

Cravings are likely a result of hormonal changes that alter taste perception. One strong argument for hormones as the culprit is that women tend also to undergo strong food cravings (and aversions) during menopause, another period when hormones are raging and changing. Janet Pope, an associate professor of nutrition and dietetics at Louisiana Tech University, told *Imponderables* that pregnant women evaluate flavor differently, so they may try different foods or combinations of foods to find foods that will now satisfy them.

And then there is the indisputable fact that the little fetus is draining some nutrition from the mother. "You are now eating for two," so the cliché goes. But most nutritionists believe the average female need consume only 300 extra calories a day of a well-balanced diet to compensate for the other life she is carrying. It would seem logical to assume that the fetus is taking in nutrients unevenly, and that is the reason for weird cravings and aversions. Ethiopian women believe that their sudden aversion to usual staples can be explained by their babies' distaste for that particular food. But biologists and nutritionists still can't explain the unpredictability in food preferences during pregnancy.

Some cravings are relatively easy to explain on a strictly nutritional basis. For example, a woman who craves olives or pickles might be low in sodium. A newfound peanut-butter fanatic might need additional protein, fat, or B vitamins. But sodium can be obtained from Triscuits or pretzels, too. Protein, fat, and B vitamins are contained in fish or meat, as well. Cross-cultural studies indicate that most mothers crave nutritious items that are *not* part of their regular diet. In the West, many expectant mothers swear off meat; where meat is prized but scarce, it is among the most common cravings. Unusual food cravings may also be, in part, an attempt to find new food combinations to stave off some of the unpleasant symptoms of pregnancy, such as morning sickness.

Almost as many women experience food aversions as strong cravings, often from foods and drinks that they enjoyed before pregnancy. One theory is that aversions are nature's way of assuring the fetus obtains good nutrition by diversifying the diet of the mother. This might explain why in Third World countries, poor women often experience aversions to staple grains—many mothers' normal diets contain too much cereal and starch, and not enough protein and fat.

Others contend that food aversions are a way of safeguarding the fetus by making dangerous substances unpalatable to the mother. Some chain-smoking, coffee-sipping, booze-guzzling females find it remarkably easy to shed their vices when pregnant. They might maintain that their sudden upgrading of habits is done out of altruism, but studies indicate that these were among the most common aversions even before their potential damage to the fetus was known. Likewise, many women find themselves nauseated at even the thought of consuming raw meat, sushi, or soft cheeses, substances that are usually safe to consume but do offer increased health risks if prepared inadequately.

But some cravings have no conceivable nutritional advantage. Perhaps the most popular craving of pregnant women is ice: *ice,* not water, a Popsicle, or a soda. Ella Lacey, a nutritionist at Southern Illinois University's medical school, says that nobody understands why women often crave foods that offer few, if any nutrients, let alone the particular nutrients she might lack. She theorizes that it may be some form of addictive behavior, where there is a drive to gain satisfaction, even if the outcome doesn't fulfill the deficiency.

The most aberrant addictive craving is pica, a condition most prevalent in the South of the United States and Central America, in which folks crave and eat non-food substances, often dirt, clay, chalk, dishwasher detergent, and, least scary, ice chips. Pregnant women comprise the largest, but by no means only, group of pica practitioners, but in most, the desire goes away once the baby is born.

Pica is more prevalent among poor folks, many of whom have nutritional deficiencies, leading some nutritionists to believe that pica, and especially geophagy (eating of dirt and clay), is a response to an iron or calcium deficiency. The more affluent woman is likely to detect such a deficiency by consulting a doctor or nutritionist, and once diagnosed, more likely to turn to spinach and liver than the backyard for a remedy.

All of a sudden, that pickle sundae is starting to sound awfully tempting.

Submitted by Angela Burgess of Los Angeles, California. Thanks also to Jerry De Duca of Montreal West, Quebec; and Steffany Aye of Lawrence, Kansas.

Why Can't We Buy Fresh Baby Corn in Markets?

You *can* buy fresh baby corn at specialty food stores, especially Asian markets, and we spoke to three wholesale produce markets that sell the stuff. But factors conspire to keep it out of your local supermarket.

The biggest problem in marketing baby corn is that there simply isn't enough demand for the little ears. Few Western recipes call for baby corn, and the family that decides to roast baby corn to accompany the barbecued hamburger is doomed to frustration, if not starvation. First of all, baby corn in its natural state requires husking, which is more time consuming and difficult than cleaning "regular" corn. Then, even if the ears are successfully de-silked, it tends to dawn on consumers that baby corn is, uh, tiny, and not sufficient to sate the appetite.

Most tend to buy baby corn as a novelty or garnish, according to Eddie Fizdale, of Peak Produce in Washington, D.C., especially as the little critters are more expensive than their full-sized brethren, and the edible portion averages only 13 percent of the weight of the ear and husk. Because of the expense, Lauren Hiltner of Babé Farm in Santa Maria, California, indicates that at this point, except for the Asian market, only high-end supermarkets tend to carry fresh baby corn.

The main customers for baby corn are restaurants and salad bars. Professional chefs know how to prepare the delicate corn without bruising or overcooking it. Salad bars use either canned or jarred baby corn to save preparation time, further diminishing the market for fresh baby corn.

Although most of the baby corn found in North America is imported from Thailand, Taiwan, and Indonesia, American farmers are increasingly trying to compete with homegrown products. Farmers can sell by-products from the baby corn, such as the husk, silk, panicle, and stem to cattle farmers for livestock feed.

Best of all, any kind of corn designed for human consumption can be grown as baby corn. Baby corn is produced from regular corn plants (and can grow alongside the other crop) and picked while very immature. Although the image of a baby six-inch-high plant, with little ears hidden in the brush is charming, in reality they grow on six-foot behemoths.

Submitted by Kathleen R. Dillon of Brooklyn, New York. Thanks also to Beth Neumeyer of Las Vegas, Nevada; and Rachel P. Wincel, via the Internet.

Why Are the Strips of Staples Designed for Office Staplers Too Long for the Space in the Stapler, Leaving Little Clumps to Clutter Drawers?

Reader Howard Labow's drawers resemble those at *Imponderables Central.* We own two handheld Swingline staplers that refuse to degrade, but we wish we could say the same for our boxes of staples. The boxes are full of scruffy odds and ends—what we call "orphans"—"stripettes" of anywhere between five to twenty-five staples—enough to keep around, but annoyingly difficult to load into the stapler.

So we called Swingline Staplers and spoke to technical-support representative Anthony Lojo, who professed astonishment that we would have any problem. The standard Swingline strip consists of 210 staples, and the stapler itself is designed to provide a half-inch cushion before the follow block (the little piece that pushes the staple forward) locks.

Suspecting a whitewash, we sought refuge with Lori Andrade, a customer-service representative at Swingline's biggest competitor in office staplers, Stanley Bostitch. But Lori reported that although half-strips are available for smaller staplers, she couldn't figure out why the 210-strip wouldn't fit in a standard office stapler from Swingline or Stanley Bostitch, and all of her companies' staplers are designed to take full or half-strips.

Frustrated, we decided to further research this Imponderable by taking a field trip to the most appropriately named office superstore we could find—Staples. As befits its name, the office superstore offers a bewildering array of staplers, but every single basic office stapler offered was designed to accept 210-strips. As advocates of the scientific method, we decided to perform empirical research. We came home, emptied the magazine of our stapler of all staples, and inserted a 210-strip of staples in our Swingline. Guess what? It entered with no problem. No orphans.

Well, not exactly. Now we were left with the staples that we had extricated from the stapler. And that led us to what we hope will win us one of those "genius" fellowships.

When do we put new staples in our staplers? When the stapler doesn't work, of course. Some of the time, we do so because the magazine is completely empty. But in our experience, this doesn't occur often. Usually, a few staples have broken apart from the rest of the strip, and are not aligned properly. A pen or nail file is used to extricate the errant staple(s), and whatever significant portion of intact strip is left inside.

But perhaps the magazine is 80 percent empty. So like the prudent motorist who doesn't wait for the fuel gauge to hit E before gassing up, we try to insert the new strip before the stapler's magazine is emptied.

Any way you attack the problem, you end up with orphans. If you take out the remnants of the old strip, they become lame-duck staples. If you try to insert the full strip, it won't fit. If you break the strip in half, at least you end up with an intact fraction of an orphan strip, but it's still an orphan.

There is only one solution to this less than earth-shattering problem. If you end up with bits and pieces of staples, toss them.

Submitted by Howard Labow of San Marcos, California.

Why Do Most Staplers Have a Setting to Bend Staples Outward?

The stapler manufacturers might have blanched when we suggested that their staplers weren't large enough to accommodate a strip of staples, but they were well armed for this Imponderable. They get asked to solve this mystery often enough that Swingline put an answer on the FAQ (frequently asked questions) section of its Web site:

> On a hand stapler you have a silver metal anvil on the [top of the] base of the stapler. This anvil is where the staple legs are formed either in an upward or outward (pinning) fashion. If your staples are forming outward, this means that the anvil has been turned. To resolve, push the button under your anvil upward. This will raise the anvil, now rotate the anvil. This will change the position of the forming slots.

Staplers were invented in the late nineteenth century, but were too expensive and labor intensive (early staplers could hold a grand total of one staple) to be widely available. Before the proliferation of staples in the early twentieth century, papers were fastened together by pins, the origin of the *pinning* term for outwardly oriented staples. Because pins were not clinched, it was relatively easy to remove them without damaging the documents they were fastening. And that's still the primary advantage of the pinning feature today—Lori Andrade, at Stanley Bostitch, said that dry cleaners are fond of pinned staples for their short-term stapling jobs, precisely because it is easier for customers to detach the staples from boxes or bags. Pinning isn't as reliable a fastener as the clinched staple, but it does make it easier to add new documents to already gathered papers.

Why have you likely never used the pinning orientation for your stapler? In our experience, the minuses outweigh the pluses for any job. Not only do pinned staples not fasten well, but the "legs" of the staple stick out, and can catch on other objects, such as your hands! If you are worried about maiming documents with the standard staple, consider some other obscure alternatives—such as paper clips or a folder.

Submitted by Jonathan Steigman of Rockville, Maryland. Thanks also to "Lollis," via the Internet; Atul Kapur, of Chatham, Ontario; Jen Braginsky, of Ottumwa, Iowa; and Eldad Ganin, via the Internet.

Why Was Charles Schulz's Comic Strip Called *Peanuts*?

Before there was *Peanuts,* there was *Li'l Folks,* Charles Schulz's cartoon produced for his hometown newspaper, the *St. Paul Pioneer Press,* starting in 1947. Fortunes are not made from selling cartoons to one newspaper, however. So Schulz pitched *Li'l Folks* to the United Features Syndicate, who was interested in the work, but not the name of Schulz's strip.

UFS perceived two possible problems. Schulz's existing title evoked the name of a defunct strip called *Little Folks* created by cartoonist Tack Knight. And there was a comic strip that was already a rousing success that United Features already distributed—*Li'l Abner.*

Who decided on the name *Peanuts*? The credit usually goes to Bill Anderson, a production manager at United Features Syndicate, who submitted *Peanuts* along with a list of nine other alternatives to the UFS brass. The appeal of *Peanuts* was obvious, since as Nat Gertler, author and webmaster of a startlingly detailed guide to *Peanuts* book collecting (http://AAUGH.com/guide/) notes:

The name *Peanuts* invoked the "peanut gallery"—the in-house audience for the then-popular *Howdy Doody* television show.

Charles Schulz not only didn't like the name change, but also objected to it throughout his career. Melissa McGann, archivist at the Charles Schulz Museum and Research Center in Santa Rosa, California, wrote to *Imponderables*:

> Schulz always disliked the name, and for the first several years of the strip's run he continually asked UFS to change the name—one of his suggestions was even "Good Ol' Charlie Brown." Up until his death, Schulz maintained that he didn't like the name *Peanuts* and wished it was something else.

In his essay on the *Peanuts* creator, cartoonist R. C. Harvey quotes Schulz to show how much the usually soft-spoken man resented the *Peanuts* title:

> "I don't even like the word," he said. "It's not a nice word. It's totally ridiculous, has no meaning, is simply confusing, and has no dignity. And I think my humor has dignity. It would have class. They [UFS] didn't know when I walked in here that here was a fanatic. Here was a kid totally dedicated to what he was going to do. And then to label something that was going to be a life's work with a name like *Peanuts* was really insulting."

Gertler points out that when Schulz first objected to the name change, UFS held the trump cards: "By the time the strip was popular enough for Schulz to have the leverage, the name was too well established." But in the media in which he had control over the name, Schulz avoided using *Peanuts* alone, as Gertler explains:

At some point during the 1960s, the opening panel of the Sunday strips (when run in their full format) started saying *Peanuts, featuring Good Ol' Charlie Brown* rather than just *Peanuts* as they had earlier. Meanwhile the TV specials rarely had *Peanuts* in their title; instead, it was "A Charlie Brown Christmas," "It's the Easter Beagle, Charlie Brown," and similar names.

In fact, we're not aware of a single animated special that even contains the name *Peanuts*—the majority of titles feature Charlie Brown, and a significant minority Charlie's untrusty companion, Snoopy.

So we are left with the irony that the iron man of comic strips, the giant who created the most popular strip in the history of comics, who made more money from cartooning than anyone, detested the title of his own creation. Schulz probably appreciated not only the royalties from foreign countries, but the knowledge that especially in places where peanuts are not an important part of the diet or had no association with children, his strip was called something else: *Rabanitos* ("little radishes") in South America, *Klein Grut* ("small fry") in the Netherlands, and the unforgettable *Snobben* ("snooty"), Sweden's rechristening of Snoopy.

Submitted by Mark Meluch of Maple Heights, Ohio.

What Happens to Sandbags After Flooding Is Over?

Most folks feel about sandbags a little similarly to how most teenagers feel about their parents: They're great to have in an emergency, but once they've provided their service, get them outta here! When flooding is feared, sandbags seem to appear from nowhere and disappear when the danger has passed.

We contacted the Wisconsin Department of Natural Resources

and found a genial soul, Gary Heinrichs, of the Bureau of Water Regulation and Zoning, who informed us

> We get a lot of letters at the Wisconsin Department of Natural Resources, but I am happy to tell you yours is the first entry in the category of Sandbag Trivia.

He was happy to share with us that in most cases, a local emergency government office, a state natural resource or environmental quality agency, or the National Guard stockpiles sandbags. Although burlap bags are still in use, plastic bags have overtaken them in popularity, as they are both cheaper and more likely to be reusable, as burlap is prone to rotting when saturated with water.

In most cases, these governmental agencies will provide individuals and communities with sandbags full of clean, washed sand for no charge. Sometimes, however, the recipients of this largesse must fill and place the sandbags themselves; even with volunteer help, the cost of labor is by far the largest expense in most sandbagging programs. When done manually, it takes two people about an hour to fill and place 100 sandbags, which will provide merely a one-foot-high pile, twenty feet long.

When the emergency subsides, the town might be full of wet sandbags that are no longer needed. All of a sudden these lifesavers now seem as attractive as dirty ashtrays after a party—but much harder to clean and remove. For example, in April 1997, the Red River Valley of North Dakota, which includes Fargo and Grand Forks, suffered its worst flood in the twentieth century, and it took four contracting companies eight weeks to remove sandbags from Fargo.

But time and cost aren't the only problem with pickup—environmental hazards exist, too, that dictate what happens to the sandbags (and the sand) used in emergencies. As Gary Heinrichs explains:

> After the floodwaters subside, sandbags along levees and other water-control structures are usually left in place. The

plastic bags will eventually decompose and the sand adds height and bulk to the levee for future flood control. Besides, while it is easy to get volunteers to fill and pile bags when the water is rising, it is very difficult to find people to empty and dispose of the mucky bags after the waters have receded.

Sandbags in populated areas and around sensitive structures like wastewater treatment plants must be removed due to health and safety dangers caused by the bags. Remember, everything that floats down the river—gas, oil, sewage, chemicals—soaks into the bags. The sand may be reused, but the bags must be properly disposed of.

Wet bags tend to attract vermin and mildew, but dry bags can be reused.

Because of the labor costs and environmental risks associated with sandbags, products have been developed to replace them. Several companies sell reusable, inflatable plastic grids or walls that can be filled from the top with sand by non-proprietary earthmoving equipment. Geocell Systems, Inc., a San Francisco–based manufacturer of the Rapid Deployment Flood Wall, claims that six laborers and one loader can build a wall 100 feet long, 4 feet wide, and 4 feet high in an hour, with only 20 percent loss of sand for the next use. While the cost of these plastic devices is far higher than a bunch of plastic or burlap sacks, they tend to be much cheaper over time—meaning that this is not only the first sandbag trivia that Heinrichs has answered, but might be one of his last.

Submitted by Sandra Koteski of Elk Grove Village, Illinois.

Why Do Rinks Use Hot Water to Resurface the Ice?

Here's the problem. Skaters, even elegant flyweights like Michelle Kwan, leave gouges that get dirty and lead to uneven residue on the ice. The more skaters there are on the ice, the more defects appear.

Our hero, of course, is Frank Zamboni, an Italian immigrant, who invented the Zamboni Ice Resurfacer to solve a problem of his own. He owned a rink, the Icehouse in Paramount, California, and realized how much time and labor was wasted with his maintenance men manually hosing and sweeping the ice—a process that took three to five men an average of an hour and a half. During hockey games, six to eight employees were required to scrape the ice between periods.

In 1942, the uneducated but highly skilled at mechanics Zamboni took a Jeep and fashioned a riding resurfacer that could automate the process. After seven years of experimentation, he crafted an early version of the Zamboni Ice Resurfacer and used it at his rink. In 1950, the most famous ice skater in the world, Sonja Henie, who won gold medals at the 1928, 1932, and 1936 Olympics, saw Zamboni's machine and wanted one for her tour. Zamboni handbuilt it and Henie showed it off on her tour—rink managers clamored for the labor-saving device, and Zamboni found himself with a new business.

The genius of the Zamboni resurfacer is that the entire operation is handled with one pass over the ice, even though four separate operations are performed:

1. A planar blade scrapes off a layer of the existing ice.
2. Scraped ice that is left on the surface is collected and put into a holding tank, about 100 cubic feet, which is the bulk of the machine.
3. Water is fed from a wash-water tank over the newly cut ice. A squeegee-like conditioner then smoothes this water over

the ice and a vacuum reclaims the water back into the tank. This does not create a new surface, but conditions the newly cut ice.

4. Clean water is then spread over the conditioned ice by a "trowel," a clean board that contains a pipe bringing in the new water, and that spreads the new water in a thin, flat film. The "new" water is held in a tank with a capacity of about 200 gallons, although about 70 to 120 gallons are used in a typical resurfacing.

The poser of this Imponderable wonders why this "clean water" spread to form the topmost sheet of ice needs to be hot. Wouldn't hot water just melt some of the ice on the surface and slow down the freezing process?

The conventional wisdom is that, compared to room temperature water, hot water creates a better bond with the existing ice. It does melt the existing ice a little more, but it fills in the cracks better. When it freezes, hot water creates a smoother temperature gradient from the top of the old ice to the top of the new surface—it "integrates" better and forms a smooth top-sheet of ice.

We spoke to Raoul Lopez, the maintenance chief at the Culver City Ice Arena in Culver City, California. His ice arena uses an Olympia resurfacer, one of Zamboni's main rivals. Lopez showed us a manual from Olympia that addresses this issue:

> For the best resurfacing results, your water supply should be 85 to 95 degrees Centigrade (180 to 200 degrees F). Hot water flows into cracks in the ice surface before cooling and freezing, and by slightly melting the ice surface before freezing, the best possible bond is formed with the existing ice. Hot water holds less oxygen than cold water and therefore produces a denser, harder ice. The hard ice does not get damaged as easily and therefore does not require the resurfacing as often or as deeply, resulting in minimal ice buildup. This

> means less time spent on ice maintenance and saves wear and tear on ice-resurfacing equipment.

We spoke to a few physicists, who confirmed that hot water would have slight advantages in resurfacing ice, but they wondered whether the small gain would really be worth the cost in heating? If scientists fielding a theoretical question wonder about costs, you can rest assured that the thought has crossed the mind of rink owners. In active rinks, resurfacing is performed every hour or two, so the cost of heating 100 gallons of hot water an hour is not inconsiderable.

And guess what? None other than Richard Zamboni, now president of his father's firm, says about his ice rink:

> We don't use hot water and we never have. Cold water works fine for us, and we never have to worry about the cost of heating it. We don't recommend hot water. When [other] customers have a problem with ice resurfacing, usually sharpening the blade fixes the problem. Occasionally, we recommend trying hot water when other solutions fail.

Despite the Olympia maintenance manual, the experienced chief of maintenance, Raoul Lopez, agrees with Zamboni. His rink doesn't use hot water except for special occasions. Although he thinks that hot water smoothes the ice out better than room temperature, hot water is just too expensive to justify the small advantages. But we contacted several NHL clubs and high-level figure-skating officials, and they confirmed that hot water rules the day in their domains.

Submitted by Michael Rzechula of Elizabethtown, Illinois.

How Did Fire Helmets Get Their Distinctive Shape?

Compared to the stories about the origins of most clothing and uniforms, the history of the fire helmet in the United States is surprisingly well documented. And even more unusual, the design of the first mass-produced fire helmet has changed remarkably little since H. T. Gratacap, a maker of fine leather luggage, invented the headgear in 1836. Gratacap, a volunteer firefighter, realized the need for more protective equipment. He named his helmet the "New Yorker," and with some modifications, it is still produced today by CairnsHelmets.

Three of the features that distinguish and define a fire helmet from other protective headgear were featured almost from the start: the badge mounted in the front of the helmet; the wide ribs that run from the top of the helmet to the brim; and wide brims, particularly the unusually long brim in the back. The Cairns brothers owned a business that manufactured badges, buttons, and insignias, and suggested mounting an identification badge to the front of the Gratacap helmet. When Gratacap retired, the Cairnses took over the helmet business, and the company, now owned by Mine Safety Appliances (MSA), is still a leading manufacturer of fire helmets.

Let's take a look at the reasons for these three design elements, and how they function today:

1. *The Front.* "The front" refers not only to the obverse of a fire helmet, but also to the badges affixed thereon. As paramilitary organizations that often worked in sites with little visibility, fire departments needed a way to identify both a firefighter's unit and his rank during emergencies—the front became the means to achieve these goals. According to Dennis Stout, product manager for fire head protection and safety at E. D. Bullard Company, a firehouse's hierarchy was expressed through the front of the helmet. Chiefs' hats would have more of a stovetop design, with much more ornamentation than the

"rank and file." The front badges also had to do with competition between different firehouses, which was not unlike the rivalry between cross-town sports enemies. As Stout explains:

> In the big urban areas, they would fight to beat each other to a fire. In the New England area, for example, there was a very heavy concentration of Scottish and Irish people in the fire service. They were barroom brawlers, and very prideful in their work. They literally would race to a fire scene. Often fights would break out over territory at the actual fires. They would fight over whose company was in control of that fire. The fronts called out the company name and station numbers.

Although competition among firefighters may not be extinct, the fronts are of use for more prosaic reasons today. Denis Ryan, product-line manager for fire helmets at Mine Safety Appliances (the company that bought Cairns in 2000), told *Imponderables* that fronts are still used to identify personnel, especially at big fires, where many different units are participating. Ryan notes that by wearing easily identifiable fronts, personnel can be held accountable for problems. Some fronts have Velcro pieces that can be removed, and placed on a board so that officers can keep track of exactly who is inside a building at any given time. And fronts can still let everyone know who is in charge:

> You can customize the front in a million ways. A regular firefighter will have a smaller front, usually six inches tall. If you're the fire chief, you'll have one that's a little bigger—maybe eight inches, and a little more elaborate.

Some fire departments use different-colored hats to indicate rank, so that a commander can better keep track of personnel. The most common scheme: firefighters and engineers wear yellow helmets; officers (e.g., captains and lieutenants) don red helmets; and chief officers sport white helmets.

2. *The Ribs.* As Stout so elegantly puts it, "It's hard to take a single piece of dead cow and shape it." Since all of the early fire helmets were fashioned from leather, Gratacap added resin to strengthen the material, and then segmented the helmets by sewing in ribs to stiffen the relatively pliant leather. In the nineteenth century, communities often relied on the expertise of local cobblers to fashion their helmets, but almost all were constructed in quarters with seams to create the domelike effect on the top. Although the ribs are no longer necessary to support modern fire helmets, most still feature them.

3. *Longer Brims in the Back.* Before Cairns Brothers mass-produced fire helmets, most of the protective headgear for firefighters featured flat brims, according to Karen Del Principe, museum manager for FASNY Museum of Firefighting, in Hudson, New York. When they would get wet, the brims would go limp. Cairns developed the curved brim, which offered a huge advantage: When water hit the curved brim, the moisture would slide right off. The New Yorker–style helmet, with a shorter front brim, was probably modeled after the helmets worn by steam engine engineers, Del Principe suggests, because it was dangerous for the engineers to hit their helmets every time they leaned forward to work on an engine.

The main purpose of the big rear brim was to keep water from running down the collar of a firefighter, according to Arnold Merkitch, retired New York firefighter and author of *Early Fire Helmets*. Merkitch notes that hot water from fires is particularly dangerous, so the back brim, which is typically three inches in length, helps protect the neck and back of the firefighter. The front brim, about half as big, assists in keeping debris off the firefighter's face, diverting most water away, and serving as a convenient place to attach protective goggles and other equipment.

Denis Ryan observes that the functional utility of the big back brim has been greatly reduced because of the increase in protective clothing. Firefighters' jackets now ride up higher on their necks. But the back brim still offers an extra measure of safety against water, debris, and even flames, especially when the firefighter is in a crouching position.

Over the past century and a half, lighter and more inexpensive materials have been introduced for fire helmets. In the 1920s, Cairns introduced aluminum helmets; in 1981, fiberglass; in 1992, Kevlar composite. What have changed surprisingly little are the styles themselves.

There is such a thing as a "modern" fire helmet, which looks not unlike a construction hardhat, and features a still visible but much diminished brim. The modern style enables firefighters to slip in and out of tight spaces, such as during auto rescue work, without the helmet getting in the way, yet still offers excellent protection against high heat and flames.

But Denis Ryan reports that the traditional shape is still popular, so helmet manufacturers offer an alternative—fire departments can buy helmets with the "old-school" look but fashioned from modern materials. The attraction is clear—plastic helmets cost about half the price of leather equivalents.

What is more surprising, perhaps, is that leather helmets are making a comeback. There is no doubt that the proponents of leather headgear are much more passionate about their choice (it's hard to get sentimental about wearing Kevlar atop your head). We heard from James L. Jester, a firefighter for the Salisbury, Maryland, Fire Department, and member of the "Salisbury Fools" (Fraternal Order of Leatherheads Society). His society's Web site speaks of the devotion to the leather helmet as an icon:

> The Leatherhead is a term used for a firefighter who uses the leather helmet for protection from the hazards we face every day on the streets. The Leather Helmet is an international sign of a Firefighter, a symbol that is not only significant in tradition from the early years of firefighting, but one of bravery, integrity, honor and pride. This helmet is a sign of who we are, not what we are.
>
> The leather helmet of choice for Salisbury Fools is the Cairns & Brother New Yorker N5A. Introduced in 1836, the New Yorker helmet has remained virtually unchanged through 166 years of faithful and steadfast service. The New Yorker helmet retains the same look and quality that generation after generation of firefighters has relied upon. . . .

Denis Ryan reports that after 9/11, firefighters from all over the country saw the New York firefighters, and there was a surge of popularity for the traditional leather helmets. Even if many of the traditional leather helmet's salient design features are no longer crucial to their protection, firefighters are slow to change equipment that has served them well for almost two centuries.

Submitted by Sroy Freedman of Lakewood, New Jersey.

Why Is Peanut Butter Sticky?

Sometimes life is not simple. Peanuts are not nuts (they are legumes). Peanut butter isn't butter (there are no milk products in peanut butter). Peanuts aren't sticky (but peanut butter is).

It's not like we would use peanut butter in lieu of Krazy Glue, but along with reader Richard Rothstein, we wonder what happens to the 850 whole shelled peanuts that go into one eighteen-ounce jar of peanut butter to up their stickiness quotient. You've probably seen peanut butter ground from fresh salted peanuts in health food stores. No other ingredients are added, and yet the end result is as sticky as the best-selling commercial brands, such as Skippy, Jif, and Peter Pan, which may contain small quantities of sugar or stabilizers (usually vegetable oil-based, to prevent the natural oil in the peanuts from separating). Although we thought that perhaps added oils were responsible for the stickiness of peanut butter, all of our sources agreed that the answer lies elsewhere.

What exactly is stickiness? When we think of a sticky substance, we tend to think of glue or masking tape, something that adheres to other stuff. But in its simplest terms, when a substance's molecules bond together in some way, we call it sticky. The stronger the bond, the stickier it is. Although peanut butter has some adhesive qualities (dip your finger into a jar of Skippy and see how well the peanut butter clings), its adhesive qualities are not strong enough to tempt us to try securing a box of books with it.

But there's another form of stickiness—viscosity. A viscous fluid resists flowing, and resists changing form when subjected to an outside force. We usually equate fluids with liquids, but thick substances such as peanut butter can be considered a fluid and possess a viscosity. Water has low viscosity and flows quickly and smoothly and can be stirred with a spoon with virtually no resistance. High-viscosity fluids, such as tar and motor oil, do not move so efficiently, and peanut butter, as you might suspect, isn't likely to flow down a drain easily. Try stirring peanut butter (as fans of "natural" peanut butter are likely to do to mix the oil that rises to the top with the rest of the "butter")—it'll strengthen your wrist.

On a plate or in our mouths, we tend to perceive high-viscosity foods as sticky, but not in the same way as the Jujubes that have clung to our teeth or the burned eggs that have stuck to an untreated frying pan. When we spread peanut butter with a knife, some of it sticks to the utensil, partly from adhesion, but mostly because adjacent molecules in the peanut butter stick together because of the food's high viscosity—the peanut butter resists flowing—it wants to stay in the same shape it was in.

Of the two main reasons why peanut butter is stickier than peanuts, both have to do with the higher viscosity of peanut butter. The first culprit is the size of the nut particles used in peanut butter. We corresponded with Harmeet S. Guraya, a research food technologist at the Agricultural Research Service, a branch of the U.S. Department of Agriculture. Guraya, whose specialty is analyzing the textures and flavors of food, wrote:

> When you eat peanuts, when the particles are bigger during initial rupture or breakup of peanut, it is easy to chew. When

you masticate the peanut and make it into a fine paste, then it gets sticky in your mouth. Commercial machines that make peanut butter grind the peanuts to different particle sizes, which is why you have different [levels of] stickiness [and thus spreadability] in different brands."

Although it may seem counterintuitive, the *smaller* the particle size, the more viscous the final product will be. This explains why if you keep chewing peanuts, it gets harder and harder to chew as you progress—you are making peanut butter in your mouth!

The second reason why peanut butter is stickier than peanuts is that the grinding of the peanuts changes the molecular structure of the legume, releasing more viscous components. Sara J. Risch, Ph.D., a food scientist and consultant to the Institute of Food Technologists, told *Imponderables*:

> In whole peanuts, the components such as the proteins and carbohydrates are enclosed in cells and held together. When the peanuts are ground, all of the cells are broken down and these materials released. This yields a material that is very viscous and thus sticky to the touch. Some of the stickiness [of the final product] could also be due to the proteins and carbohydrates. It is not from the oil.

Guraya emphasizes that viscosity is not the same thing as stickiness, and that

> the sensory perception of stickiness could be due to a variety of reasons for different foods. When I gave you the reasons for stickiness of peanut butter, it pertains only to peanut butter, although the definition of stickiness stays the same.

Although a few other factors, such as the type of stabilizers and the inclusion of high-moisture peanuts, can up the viscosity quotient of peanut butter, the grinding process itself explains the perception of stickiness. Americans must not mind coating the roofs of their mouths with the stuff—just short of one-half of the entire U.S. peanut crop is used to make peanut butter.

Submitted by Richard Rothstein of Haddonfield, New Jersey.

Why Are the Uniforms of Professional Japanese Baseball Players Printed in English Letters and Arabic Numbers?

Japanese society is sometimes accused of insularity, yet the country has not only adopted American baseball as its favorite team sport, but displays the names of its heroes in English. How did this happen?

Most historians date the birth of baseball in Japan to the early 1870s, at what is now Tokyo University. Horace Wilson, an American professor, taught students how to play. According to Ritomo Tsunashima, an editor and writer who writes a column about uniforms for *Shukan Baseball* magazine in Japan, these students were probably then wearing kimonos and hakama (trousers designed to be worn with kimonos).

In 1878, the Shinbashi Athletic Club established the first baseball team in Japan, and they were the first team to wear a uniform. The team's founder, Hiroshi Hiraoka, was raised in Boston, Massachusetts, and bought the equipment and uniforms to conform to what he saw in the United States at that time. Tsunashima observes that in the early days of the game in Japan,

> The players of baseball were fans of the Western style of life, and they were fond of Western culture. When exactly alphabet letters appeared on uniforms is unknown, but likely it was shortly after letters appeared on uniforms in the United States.

The first professional team in Japan that had uniforms with printed English words and names was founded in 1921, but disbanded in 1929. But in 1934, a U.S. all-star team containing such legendary figures as Babe Ruth and Lou Gehrig visited Japan. To challenge the visitors, Japan assembled a team that eventually became the Yomiuri Giants. When the Giants returned the favor the following year and visited the United States, their uniforms contained the kanji (i.e., Japanese) alphabet and numbers, but this was an anomaly.

When the first professional baseball league formed in Japan in 1936, the players appeared with an English alphabet and Arabic numbers on their front—the Tokyo Giants appeared with "Giants" emblazoned on their chests. During World War II, kanji characters reappeared, as Western symbols were obviously frowned upon. Tsunashima writes:

> Professional baseball continued during the bombing raids by the United States in 1943, but playing baseball was finally forbidden in 1944. After World War II, professional baseball started again with the English language [on the uniforms], and kanji letters were never used again.

Why has English superseded kanji, especially when amateur and school baseball teams usually use Japanese characters? We haven't been able to find any answer more fitting than Robert Fitts's, owner of RobsJapaneseCards.com:

> Because it's cool. Just the same as when people call me and say, "I want Japanese baseball cards written in Japanese," and I have to tell them, "They don't make them."

Submitted by Marshall Berke and Chris Tancredi of Brooklyn, New York.

Why Does Caucasian Babies' and Children's Hair Get Darker as They Age?

Our hair color is determined by genetics, but in some cases Mother Nature chooses to not reveal our ultimate hair color until well into adolescence. During infancy, the melanocytes, skin cells that mark and deposit pigment, are not fully active, and don't function in many children until sometime during adolescence.

Scientists still don't understand exactly why hair darkening occurs in fits and starts throughout childhood and adolescence, and differs so radically from person to person. Dermatologist Joseph Bark, author of *Skin Secrets: A Complete Guide to Skin Care for the Entire Family*, wrote *Imponderables* that the eventual darkening of hair color seems to be a

> slow maturation process rather than a hormonally controlled process associated with the "juices of puberty," which causes so much else to happen to the skin of kids.

As is often the case with medical questions that are curious but have no practical application (or grant money attached), the definitive solution to this Imponderable is likely to remain elusive. As Chesapeake, Virginia, dermatologist Samuel Selden celebrates:

> I don't believe that much study has been made of this, and until that is done, it means that armchair speculators like myself can have a field day with answering questions like this.

Submitted by Debra Allen of Wichita Falls, Texas. Thanks also to Chester A. Tumidajewicz of Amsterdam, New York; Tina Litsey of Raceland, Louisiana; and Edward Litherland of Rock Island, Illinois.

Why Does the Bass Clef on Sheet Music Have Two Dots Next to It? What's the Purpose of the Fussy Look of All the Clefs? And How Did Musicians Share Compositions Before There Was Sheet Music?

We hope that trained musicians will bear with us while we cover (and simplify) a few basics so that even we music illiterati can understand the answer to this Imponderable. Even those of us who don't play music are familiar with the musical staff, which looks something like this:

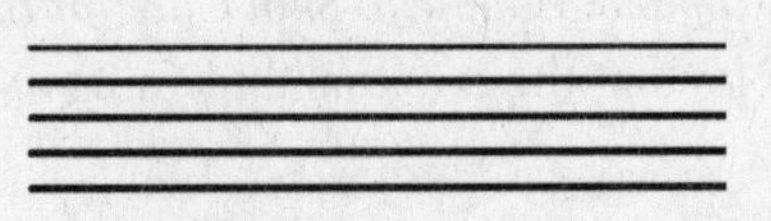

By the arrangement of notes on this staff, a musician can tell how high or low the notes should be (the pitch), when each note should start (temporal position), and how long each note should last (duration). The notes are named after the first seven letters of the Roman alphabet, starting with A on the bottom until G is reached. Notes may be placed on a line, in the space between lines, or above the top line or below the bottom line of the staff.

If instruments had a range of notes that could be encompassed by the five-line staff, life would be much easier for composers and musicians. But there are many more notes on a piano, for example, than can be annotated within a five-line staff. One way of simplifying this problem is by inserting a clef at the far left side of the staff (all sheet music is read from left to right, even in Israel). The clef's job is to inform musicians what pitch the first note of the piece will be played in, and therefore determining the pitch of all the other notes on that staff in relationship to the clef. To answer the second part of this Imponderable first, the fancy curlicues on the clefs are not just for effect—they point out the pitch of the first note on the staff.

Probably the most familiar clef is the treble clef, also known as the G clef. The end of the inside curlicue (the fine tip that looks like a tail) marks the line, usually the second from bottom line, where the G note is designated:

Look at the G clef again and you'll see that the entire clef is a stylized, cursive rendering of the letter *G*.

Lesser known is the alto clef, where the middle C note is marked by the center of two stylized backward *C*s:

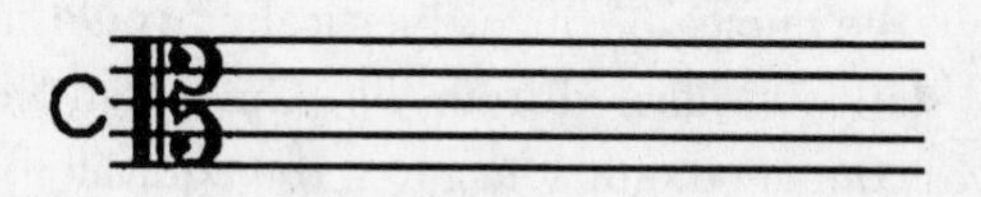

The middle C would lie on the middle line; D would be in the space between the C line and the line above it (the E line).

And now we get to the star of our Imponderable, the bass clef. As you've probably guessed, the bass clef also marks a line, in this case the F below middle C. On a piano, the left hand is usually following the instructions on a staff headed by a bass clef, while the right hand is written in treble clef. The bass clef looks like this:

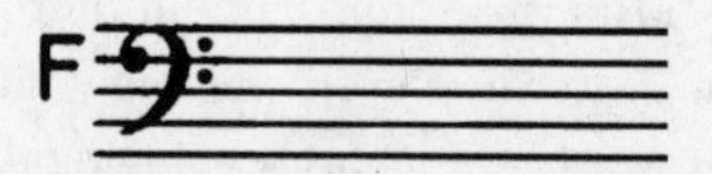

Those two dots aren't to the right of the clef—they are actually part of the clef and essential for two reasons. Just as the treble clef is the rendering of a cursive letter *G,* the bass clef is a rendering of the letter *F.* Those two dots are a stylized expression of the two limbs coming off the base of the *F.* And just like the G and C clefs, the stylization has a purpose, as Michael Blakeslee, of the Music Educators National Conference puts it: "To help the reader zero in on the note defined by the clef, the dots appear above and below the line for the F note."

Our musical staves are the descendants of a Benedictine monk, Guido d'Arezzo, who lived in the early eleventh century and composed chants. He faced the problem of how to teach his monks new chants before the advent of standardized musical notation, and an even more daunting task of imparting his new music to priests in far-flung monasteries. D'Arezzo developed the notion of using his hand as a way of dictating the pitch of his chants. He held his left hand with his palm facing away from him, fingers in horizontal position, and pointed to the four fingers (the thumb wasn't long enough to provide enough information). The knuckles and joints would indicate timing (horizontally), the choice of finger (vertically) would indicate pitch. Our musical staff was adapted from the "Guidonian hand" (the first musical staves contained only four lines, the equivalent of d'Arezzo's left forefinger, middle finger, ring finger, and pinkie), which used a six-note scale.

D'Arezzo continued to think of ways to standardize musical notation. One of his great contributions was the solfeggio, a group of six syllables designed to serve as a mnemonic to correspond to the notes in his system. Five of them will sound familiar: "ut re mi fa sol la." Why wasn't it "*do* re mi"? These syllables were not nonsense sounds, but rather from the first syllables of Latin phrases in a well-known chant in honor of St. John, "Ut Queant Laxis." ("*Ut* queant laxis," "*Re*sonare fibris," "*Mi*ra gestorum," *Fa*muli tuorum," "*So*lve polluti," and "*La*bii reatum."). Because everyone knew this chant, the solfeggio became the answer to two choruses singing the same note at the same pitch without the benefit of musical notation.

Why was ut replaced by do? We'd like to think it was to make Rodgers and Hammerstein's life a little easier (it's easier for Julie Andrews to sing a song that starts with a reference to a female deer than to an ut!), but history records a more prosaic answer. Another Italian, Giovanni Maria Bonocini, suggested the retirement of ut in 1673, because of the more open, mellifluous sound of do compared to the closed, harsh ut. And probably at least as important, do represents *Dominus,* the Latin word for "Lord."

The difficulties in developing a universal system of musical notation were exacerbated by the difficulties of disseminating scores. Even after the invention of the press, printing sheet music was a technical nightmare. Today, with computer programs and photocopying, the costs are relatively trivial, even if the music isn't necessarily any better than it was centuries ago.

Submitted by David Ward of Oakville, Ontario. Thanks also to James King of Honolulu, Hawaii; and Donald Ullrich of Burlington, Iowa.

Why Can't You See Stars in the Background in Photos or Live Shots of Astronauts in Space?

There actually are folks out there who believe that NASA pulled off a giant hoax with the "so-called moon landings." Often, the lack of stars in the background of photos of the astronauts is cited as startling evidence to support the conspiracy.

Sheesh, guys. If you want to be skeptical about something, be dubious about whether "When you're here, you're family" at Olive Garden, or whether State Farm Insurance will be there for you the next time you're in trouble. But don't use a dark background in a photo of outer space to convince yourself that astronauts have never gotten farther into space than a Hollywood soundstage.

The answer to this Imponderable has more to do with photography than astronomy. Next time you go to a football game on a starry night, try taking a photo of the sky with your trusty 35mm point-and-shoot camera or camcorder. Guess what? The background will be dark—no stars will appear, let alone twinkle, in the background.

The stars don't show up because their light is so dim that they don't produce enough light on film in the short exposures used to take conventional pictures. But you *have* seen many photos of stars, haven't you? These were undoubtedly time-lapse photographs, taken with fast film and with the camera shutter left open for at least ten to fifteen seconds. Without special film and a long exposure time, the camera lens can't focus enough light on the film for the image to appear. Jim McDade, director of space technology for the University of Alabama at Birmingham, elaborates:

> Even if you attempt to take pictures of stars on the "dark side" of the Earth during an EVA [an extra-vehicular activity involving astronauts leaving the primary space module, such as a spacewalk] in low-earth orbit, a time exposure from a stable platform of about twenty seconds is necessary in order to capture enough stellar photons to obtain an image showing stars, even when using fast films designed for low-light photography.

The same problems occur with digital cameras, film, and video cameras, as McDade explains:

> A digital camera, a film camera, and the human eye all suffer similar adaptability problems when it comes to capturing dimmer background objects such as stars hanging behind a space-walking astronaut in the foreground. The human eye is still much more sensitive than the finest digital or film camera.
>
> Photographic film is incapable of capturing the "very bright" and the "very dim" in the same exposure. The lunar surface is brilliant in daylight. The photos taken by the Apollo astronauts used exposure times of a tiny fraction of one second. The stars in the sky are so dim that in order to capture

them on film, it requires an exposure time hundreds of times longer than those made by the Apollo astronauts.

Those of us who live in the city have had the experience of going out to a rural area on a clear night, and being amazed at the number of stars we can see when there aren't lights all around us on the ground. You can create the same effect inside your house. On a clear night, kill the lights in a room and look out the window. Depending upon the atmospheric conditions, a star-filled sky may be visible to you. Flip on the lights inside the room, look outside, and the stars have disappeared.

Why? Light from a bright object near us can easily dwarf light emanating from distant objects, such as stars. In the case of astronauts, the lights attached to the space vehicle or space station or even the lights on an astronaut's helmet can wash out the relatively dim light from the stars in the background.

Even with sensitive film, the suits that astronauts and cosmonauts wear reflect a lot of light. The glare from the astronauts themselves will provide contrast from the dark sky background and faint stars. Any light emanating from the stars is unlikely to be exhibited when cameras are geared toward capturing clear shots of a space walker.

Perhaps the space conspiracists would stifle themselves if the Apollo astronauts had taken time-exposure photographs that could display the stars in all their glory, but they never did. As McBain puts it: "After all, they went to the moon to explore the moon, not to stargaze."

Submitted by Scott Cooley of Frisco, Texas. (For much more information on the issues of photography in outer space, see Jim McDade's "Moonshot" Web site at http://www.business.uab.edu/-cache/Defaultb.htm.)

What Are Those Wigwams Doing off to the Side of Interstate Highways?

As reader Ann McGinnis-O'Connell suspects, these aren't homes of Native Americans, but rather storage bins for salt (and sometimes sand), for use by public works agencies in de-icing roadways. They are located so close to the highway because the salt needs to be picked up as quickly as possible and distributed onto nearby roadways. Most of the time, the storage structures are on land too close to the highway to be of value for residential or commercial use.

Why is it necessary to store the salt in buildings, anyway? Just like your table salt, if de-icing salt is exposed to moisture (not just rain, but high humidity), it can cake up or get lumpy, and be difficult to spread. Water, whether in the form of rain or streams, can contaminate or dissolve it, too.

The storage domes are larger than they may appear as you whiz by them in your cars. One large manufacturer, Dome Corporation of America, manufactures structures ranging from 40 to 150 feet in diameter, the latter behemoth holding nearly 20,000 tons of salt and more than 26,000 tons of sand (sand is denser than salt, accounting for the weight differential). According to the Salt Institute, these humble storage buildings are designed to withstand heavy snows and winds of up to eighty miles per hour.

Why the upside-down ice-cream-cone shape? One requirement for a storage structure is that it not contain posts or beams that get in the way of loading or unloading the salt. In some cases, salt or sand are poured into the structure from the top, but most transportation agencies want the option to push the stuff in from the front, and unload it by driving trucks inside the structure—the bottom-heavy conical shape permits a stable, freestanding building.

But you could also fit a truck or forklift into a rectangular building, although it might be more difficult to manufacture such a structure without beams or trusses. The real secret to the conical shape is that it wastes the least amount of space. Richard Hanneman, president of the Salt Institute, points out that if you pour massive amounts of salt into a structure from an opening on the top, the salt naturally creates a cone shape. Salt and sand, like all loose, unconsolidated granular material, has an "angle of repose," the steepest angle at which the mound will rest without slippage. If a forklift attempts to steepen the angle, say by pressing the salt on the bottom toward the middle of the cone in order to load more salt, it disturbs the equilibrium of the mound, causing grains to fall and creating pressure against the walls of the structure. In nature, ocean tides can similarly destabilize land near the shore by eroding the base, steepening its angle beyond its natural repose: Landslides are the result.

As Hanneman points out, "If you load the salt from the top, it will come down in a conical shape," regardless of the type of structure that contained the salt. The angle of repose for salt is thirty-two degrees, and most salt domes are built with a similar slope, thus minimizing the size, and the cost, of the "wigwam."

There is something charming about these wigwam structures, which are often covered with red shingles, but are otherwise unmarked and unadorned. But they might have some tough competition from a newfangled storage structure that looks positively old-fangled—rectangular buildings that look like barns. We were particularly taken with the "Hi-Arch Gambrel" (a gambrel is a roof with two slopes on each of two sides, with the lower slope steeper than the upper) produced by Advanced Storage Technology of Elmira, New York. We chatted with representative Jane Kerber, who remembers going on road trips in the 1960s and asking her parents, "What are those?" in reference to the salt bins.

Understandably, Kerber is high on the merits of her company's thirty-feet-high structures. Domes have a large clearance in the center, but not near the walls. Trucks can't always load into the front of a dome because of a lack of clearance, so are forced to load from the top, which often requires expensive conveyors and necessitates dumping the salt or sand on the ground outside of the dome before it is put on the conveyor. The Hi-Arch Gambrel is large enough so that trucks can enter the structure and dump directly inside, raising their beds inside the building.

Many communities mix salt with "stretchers," such as sand, grit, or chemical anti-skid concoctions. These other abrasives are cheaper than salt, and they also provide traction for vehicles when it is too cold even for salt to help melt ice until the temperature rises. Advanced Storage Technology argues that it is easier to mix these substances in its "barns," since there is much more headroom, especially as you get near the walls.

Kerber also makes the aesthetic argument against the "wigwam":

> People driving along seeing our building might not even recognize that its design is for salt storage. It looks like a barn. This is an advantage, aesthetically. People recognize that the salt-storage domes have an industrial use.

Who would have ever thought that folks in the salt-storage business would be interested in aesthetics? You can judge for yourself by looking at the "wigwams" (http://www.dome-corp-na.com) and barns (http://www.saltstorage.com) on the Web.

Submitted by Ann McGinnis-O'Connell of Chicago, Illinois.

Why Don't Women Faint as Much as They Used To?

The most common cause of fainting is a lack of sufficient blood flow to the brain because of a sudden drop in blood pressure. Serious heart problems, including arrhythmia and ventricular tachycardia, can also be the culprit and less frequently, neurological irregularities. Some phobics truly do faint at the sight of blood or a coiling snake, but this is the result of a sudden loss of blood pressure.

None of these medical conditions has been eradicated, so it's always a shock to pick up a Victorian novel and find damsels fainting when they are frightened, fainting when they are ecstatic, fainting when their heart is heavy, and most of all, fainting when it is convenient to their purposes. What ever happened to swooning?

The most often cited faint-inducer in the past was a torture device known as the corset, an undergarment so tight that according to Lynda Stretton's essay "A Mini-History of the Corset,"

> By the time they were teenagers, the girls were unable to sit or stand for any length of time without the aid of a heavy can-

vas corset reinforced with whalebone or steel. The corset deformed the internal organs, making it impossible to draw deep breath, in or out of a corset. Because of this, Victorian women were always fainting and getting the vapours.

What was the purpose of this torture? The mark of a beautiful woman was thought to be the thinnest possible waist. Stretton reports that although the literature refers to corseted waists as small as twelve to eighteen inches in adult women, most of these figures were probably fantasies:

> Measurements of corsets in museum collections indicate that most corsets of the period 1860 to 1910 measured from twenty to twenty-two inches. Furthermore, those sizes do not indicate how tightly the corsets were laced. They could easily have been laced out by several inches, and probably were, because it was prestigious to buy small corsets.

Even so, trying to cinch in a waist six or eight inches tighter than nature intended could do damage to the circulatory system.

Corsets, with certain Madonna-like exceptions, are no longer a fashion rage. Has fainting stopped? Nope. As Louis E. Rentz, executive director of the American College of Neuropsychiatrists, wrote us:

> [Fainting] is one of the most common complaints that present to a physician's office and certainly one of the most common things neurologists and cardiologists see. It is common at all ages.

Two U.S. studies of more than 5,000 healthy people indicate that somewhere between 3 to 6 percent of the population report at least one fainting episode over an extended period of time (ten years in one study, twenty-six in another).

One statistic popped out on all of the research about fainting we consulted: Men fainted almost as much as women. Could all the reports of female swooning in the nineteenth century be inaccurate? Exaggerated?

Upper-crust society in England, the subject of most of the Victorian novels we read in English 101, were living in times when women were considered to be delicate creatures. Corsets were thought not only to make women look more attractive, but to help them medically (children as little as three or four wore corsets, in the mistaken belief that such support would help strengthen their posture and musculature).

Even more bizarre were the beliefs about the "delicate sensibilities" of ladies. The Society for the Reformation of Manners, founded in London in 1690, first started fighting against prostitution and drunkenness, but by the nineteenth century became preoccupied with "cleaning up" the language. *Bloody* became a taboo word, and no host would serve a woman a rare cut of beef for fear that the sight of blood would send a delicate lady to the fainting couch for a swoon. In her review of etiquette books of the period, *The Best Behavior,* Esther B. Aresty notes that not only sticks and stones but words were believed to be able to hurt Victorian females:

> Well-bred English people never spoke of going to bed; they retired. Even a bureau could not have "drawers." To refer to a female as a "woman" was insulting and a foreigner might cause a fainting spell if he said "woman" to a lady's face. . . . Delicacy was, in fact, being carried to such extremes that Lady Gough's *Etiquette* ruled that even *books* by male and female authors should "be properly separated on bookshelves. Their proximity unless they happen to be married should not be tolerated."

Women in the nineteenth century lived in a culture in which fainting was seen as a badge of femininity. When Alexis de Tocqueville came to the United States as a young man in the 1830s, he observed that European women were:

> seductive and imperfect beings . . . [who] almost consider it a privilege that they are entitled to show themselves futile, feeble, and timid. . . . The women of America claim no such privileges.

In this environment, who could blame a woman for timing a swoon so that it coincided with the approach of her intended? If you wanted to avoid a nasty confrontation, why not faint instead?

Our theory is that fainting was largely a cultural phenomenon, a benign form of mass hysteria. No doubt, tight corsets constricted blood flow and caused fainting, especially in women with low blood pressure. But Victorian society rewarded fainting—it was considered feminine and attractive behavior. The faint became an all-purpose excuse for ducking difficult obligations, the nineteenth-century equivalent of "my dog ate my homework," with the added benefit of garnering sympathy.

We bet that the actual incidence of fainting hasn't changed that much over centuries. Indeed, we recently encountered a reality show that proved that not only Victorian damsels feign fainting. On Fox TV's reality show, *Boot Camp,* a drill instructor found macho Meyer dawdling on a hot day. Meyer decided to evade doing his required push-ups by "fainting." He didn't fool his comrades, just as we suspect most swooning damsels didn't. But maybe Meyer's fate explains why you don't see folks bragging about fainting anymore: he was unanimously booted off the show.

Submitted by Nathan Trask of Herrin, Illinois.

Why Don't Tornadoes Ever Seem to Hit Big Buildings or Big Cities?

The operative word in this mystery is *seem*. For all of our meteorological experts agree with Harold Brooks, the head of the Mesoscale Applications Group of the National Severe Storms Laboratory in Norman, Oklahoma, who states:

> There is a myth that tornadoes don't hit downtowns, but that is just a myth that comes from the fact that downtowns are small areas. If you randomly picked any other similarly sized areas in the middle of the United States, they wouldn't get hit often, either.

In other words, tornadoes are non-discriminatory offenders, and are subject to the laws of probability. The land covered by major population centers is tiny compared to the total expanse of North America, but a big city is theoretically just as likely to get hit as all the local trailer parks if they covered as much of an expanse as the downtown area.

There are portions of the United States where tornadoes are much more likely to hit, however. Tornadoes have been tracked in all fifty U.S. states, but the so-called tornado alleys are the Midwestern area from Texas all the way north to the Canadian border, in the southeastern United States, and in the Ohio Valley and southern Great Lakes region, extending as far east as western Pennsylvania. But within that general area, trailer parks, skyscrapers, the Mississippi River, or the Dallas Cowboy cheerleaders can't stop tornadoes.

One other reason why the perception may have spread that tornadoes don't hit big cities is that many of the densest population centers, such as the Washington, D.C.–to–Boston and the San Diego–to–San Francisco corridors, are outside of "tornado alley." Even so, tornadoes *do* sock big cities. Chuck Doswell, senior research scientist at the Cooperative Institute for Mesoscale Meteorological Studies, wrote us:

> Tornadoes have recently hit the downtown areas of such cities as Salt Lake City, Utah; Miami, Florida; Nashville, Tennessee; and Fort Worth, Texas. In the Texas event, there were at least two tall buildings hit. One suffered enough damage that it was decided to demolish it rather than to repair the damage.
>
> In 1970, in Lubbock, Texas, a violent tornado hit a large building with sufficient force to twist the structure. The only reason tall buildings are hit infrequently is that they don't occupy very much space in this nation of ours. The more area covered, the greater the likelihood of being hit by a tornado.

Dr. T. Theodore Fujita, a tornado scientist who developed the scale commonly used to rank the severity of tornadoes, considered the role of skyscrapers and population density in thwarting the development of small tornadoes. The University of Chicago professor, who died in 1998, noted that since 1921, "practically no tornadoes occurred or moved across the central portion of Chicago." Fujita theorized that perhaps the city's higher temperature than surrounding areas (a phenomenon we discussed in our first book, *Imponderables*) and its man-made structures might be "acting against any tornado activity over the city." Other major population centers that have been studied for tornado patterns, London and Tokyo, also seem to enjoy a relative dearth of small-tornado activity. But no expert seems to seriously think that even the highest skyscraper in the largest metropolis would scuttle an intense tornado from unleashing its fury.

Submitted by John Beton of Chicago, Illinois. Thanks also to Laura Gunn of Ames, Iowa.

Why Do Only Kids Seem to Get Head Lice?

Are there really insects so vile, so vicious, that they attack only innocent children? Maybe there are, but head lice are equal-opportunity pests. Anyone, young or old, white or black, male or female, is susceptible.

But that doesn't mean that head lice don't have preferences. According to the research of entomologists at Clemson University, every year 6 to 10 million people in the United States have head lice, with approximately three-quarters of them schoolchildren younger than twelve years old. If you're a head louse, your target demographic is three to eleven. This is true all over the world: Recent studies in France, Denmark, Atlanta, Israel, and Korea all found that at least 3 percent of elementary school children were infected by lice or nits, some at twice that rate.

Lice aren't demanding creatures. They just want to suck a little of your blood. All three varieties of lice (head, body, and "public") feed on mammals, but head lice are found only on humans. A louse has six legs and is incapable of flying or jumping, so they cling to human hair and bite through the scalp, creating a tiny opening to suck the blood. Without a blood host, head lice are likely to survive for only a day or two.

As if having a louse on your head wasn't lousy enough, the females have the nasty habit of laying eggs, or nits, in our hair, usually near the scalp, at a rate of about six to ten per day. Lice manufacture an insoluble cement that makes the eggs cling to our hair, so standard shampooing will not get rid of the nits. The egg hatches in about eight to ten days, and becomes an adult in another week to ten days. The adults only live for a month or so, but considering the rate at which they manufacture nits, that is of little solace.

We hope we will not be accused of nitpicking when we point out that lice eggs are often confused with dandruff or dried-up hairspray, as they are a similar color and size. But while dandruff and caked hairspray will brush out easily, nits usually can't be extricated even if you go over it with a fine-toothed comb, although the closer the teeth in a comb are, the more likely you are to rid yourself of lice and nits.

While body lice tend to invade folks with poor hygiene, and "public" lice (the euphemism for "pubic lice" or "crabs") are often sexually transmitted, there is no reason for social stigma about having head lice. Head lice are usually spread from person to person by direct contact, but they can also move to an object that has been used by an infected person, and transfer to another.

Head lice are perfectly content to reside on graybeards if given the chance, but kids provide more opportunities. Santa Monica, California, dermatologist Joseph W. Landau has seen several epidemics of lice in schools. Kids are less likely than adults to know whether they've been infested, and one kid can spread it to scores of others before treatment is provided.

More important, kids in school are constantly playing with one another, roughhousing, and touching. Girls tend to have more head lice than boys, probably because close contact is more socially acceptable, and they tend to share hats, combs, and hair accessories. Slumber parties can involve sharing more than s'mores.

Other perfectly innocent behaviors can cause louse problems.

Kids are just the right height to spread lice in chairs, whether in a movie theater, a restaurant, an airplane, or most commonly, a school bus—their head hits the seat back, and the fleeing louse parks itself in the upholstery. The next person who sits in the chair is likely to find a new blood-sucking pal. Lying on a rug after an infected person has occupied the space makes you an equally compelling target. Parents of infected kids are just as susceptible to attracting head lice, and sometimes they succumb, but the parents are likely to take preventative measures, such as medicated shampoos, once they know their offspring has lice. If they don't know about their progeny's condition, and happen to use a towel with which their infected child dried his or her hair, they will be lousy with lice.

Although head lice are annoying, they are not particularly dangerous. In fact, the harshness of some of the remedies can be worse than the itching and occasional resultant rashes and infections that arise from scratching. Probably because of the social stigma of other kinds of lice, schools sometimes overreact when there is an infestation, and send parents a warning when even one child is known to be infected. Richard J. Pollack, of the Harvard School of Public Health, rails against the prevalent "no-nits" policy of many schools in which infected children are automatically sent home until nit-free. Pollack's experience with samples sent in for diagnosis is that even among those that contained real louse eggs, "many were comprised solely of hatched or dead eggs; thus no treatment would be warranted." In a few cases, courts have ordered child custody be taken away from parents who failed to eliminate infestations.

Back in the day when a certain master of Imponderability was in grade school, about the worst thing that your enemy could say about you was that you had "cooties." Perhaps we are no more enlightened today.

Submitted by James Gleick of Garrison, New York.

Why Are Wells Round?

Next time you go swimming in a pool, please note that large quantities of water are quite capable of being held in rectangular receptacles, square ones, and even trapezoidal ones. But if you try *constructing* a trapezoidal pool, you'll wish you were building a round one. Most wells are round because fashioning a round well is easier than building other shapes. And in this case, easier means cheaper. Kevin McCray, of the National Ground Water Association, wrote to *Imponderables*:

> Water wells are constructed by means of specially designed machines or rigs that drill into the earth to encounter productive water-bearing formations. Some drilling machines work by lifting, turning, and dropping a long chisel-shaped tool, which forms a round hole by cutting up the rock or other earth materials. Another type of water well drill rotates a bit fixed to the lower end of a steel pipe called the drill pipe or rod.

Round wells have an illustrious history, dating back to the Chinese about 4,000 years ago, and to some extent, modern construction methods are just an automated version of what workers did by hand millennia ago, as McCray explains:

> The Chinese developed a drilling method now known as the cable tool percussion method. Using bamboo tools, they were able to drill wells of up to 3,000 feet deep, although the deepest holes sometimes took two or three generations to complete. Cable tool drilling operates by repeatedly lifting and dropping a heavy string of drilling tools into the borehole, turning slightly with each stroke.
>
> While cable tool drilling is still widely practiced, probably most boreholes are constructed by use of the direct rotary drilling method. The borehole is drilled by rotating a bit fixed

to the lower end of a drill rod. Cuttings of rock and soil are removed from the borehole by continuous circulation of a drilling fluid as the bit penetrates the formation.

The turning action of either the cable tool method or direct rotary method assures that a round, straight hole will be constructed. While it might be possible to push or pound a square hole into the ground, the method would be limited by its impracticality. Virtually all water wells constructed in developed nations are drilled, not dug.

The occasional square well can be found, as McCray implies, mostly in undeveloped nations. Wells have a nasty habit of caving in. The best defense against a cave-in is lining the inside of the well. Shifting of the earth, severe weather, or earthquakes can cause cave-ins, but at no time is the danger greater than during the construction of the well itself. The only obvious reason to construct a square or rectangular well, unless there are space or aesthetic considerations, is because of a lack of lining materials that can be curved to fit a round well. In poor countries, flat wood boards are sometimes used for lining, despite wood's obvious inappropriateness for holding water, and anyone poor enough to use untreated wood to line a well is unlikely to have the tools necessary to form the lining to fit a round well.

Submitted by Kathyrn Bennett of Yarmouth, Maine.

Why Are Some People Double-Jointed?

Here at *Imponderables Central,* the entire staff is unable to touch our toes without bending at our knobby knees. How can the human pretzels that perform for Cirque du Soleil place their legs behind their heads and play "Chopsticks" with their toes?

Medical authorities are quick to assert that flexible people have no more joints than mere mortals. As Timothy N. Taft, director of sports medicine at the University of North Carolina at Chapel Hill, wrote us:

> The increased mobility is caused by laxity or looseness of the ligaments and the capsule that normally provide stability to the joint. The supporting structures that normally limit joint motion are elongated or more flexible than usual, thus the increased motion—usually in an unusual direction. This laxity can occur as a result of congenital looseness or it may follow an injury.

Taft portrays hypermobility as more of a curse than blessing; orthopedists are far more likely to see patients complaining about this condition than bragging about their flexibility. Several genetic conditions lead to hypermobility, most commonly Ehlers-Danlos syndrome, an inherited condition in which the collagen (the protein that reinforces the structure of the joints) is hyperelastic. Those with Ehlers-Danlos syndrome often have other unpleasant side symptoms, including joint pain, osteoarthritis, and stretchable skin—the so-called rubber men in circus sideshows usually have Ehlers-Danlos.

Collagen is the key component in cartilage, which provides padding between bones. The more flexible your cartilage is, the more flexible you will tend to be. Children's cartilage is especially elastic, which is why pre-adolescent boys and girls are capable of twisting and folding their limbs in gymnastics; as we age and hormones kick in, the cartilage hardens along with bones, and most of us lose flexibility.

The majority of people that we might offhandedly call "double-jointed" do not suffer from any medical condition. They most likely possess supple collagen and proportionately longer discs than the vertebrae they separate. Shallow sockets, smoother bone ends, and especially flexible ligaments also can contribute to hypermobility. Needless to say, regimens of stretching will enhance a predisposition toward flexibility.

Even if "double-jointed" is a misnomer anatomically, some partisans have embraced the term. The Web site of the Halston Gymnastic Club has a "Double Jointed Page" (http://www.hgc.mcmail.com/) that argues that the *Oxford English Dictionary* endorses the concept (the fifth definition of "double" in the *Shorter Oxford English Dictionary* is: "Fold . . . so as to bring the two parts into contact, parallel; bend (the body), especially into a stooping or curled-up posture. . . ."). The site disdains the use of "hypermobility" because the prefix "hyper" is used to signify "too much."

Ironically, while the general population is striving to become more flexible, the "double-jointed" often don't take advantage of their attributes:

> Joint problems caused by not exercising joints properly are the scourge of the general population, and much emphasis is placed on the therapeutic benefits of stretching exercises for everybody. Those with unusually good flexibility will gain little from these routines because the exercises designed for the average person will not provide the range of movement to exercise their joints through their natural range. Whatever your natural range of joint flexibility, if you do not use it, the joint will degenerate.

To folks with serious genetic hypermobility, extreme flexibility isn't necessarily a gift, as their full range of motion sometimes sends their bones and ligaments to places that joints were not meant to support, while others derive a paycheck from their "double-jointedness."

Submitted by Debra Allen of Wichita Falls, Texas.

Why Do You Sometimes Find Ice in the Urinals in Men's Bathrooms?

We were a little surprised when a woman posed this Imponderable, but Judith Dahlman came prepared with a rationalization:

> In my waitress days, the restaurant where I worked had a big bin of ice right at the bar, where the waiters scooped the ice for drinks, as set-ups for the bartender to pour the booze or to fill up their own glasses with soft drinks. A metal scoop was included and you were supposed to use it, but waiters in a hurry often scooped the ice with the glass itself. This inevitably led to broken glasses, obviously a health hazard, what with pieces of glass now amid the ice cubes. Waiters were instructed to immediately lay a bar rag over the ice, as a signal to the other waiters not to use the ice. The busboy would then empty the bin, which would be washed out with water. New ice would be put in the bin.
>
> My husband Paul's response was: "So that's why you sometimes see a big load of ice in a urinal?" I admit that I have only rarely seen this phenomenon, probably because most busboys are male, and they have free access to the men's rest rooms.
>
> In my waiter days, I never thought of where the busboy dumped the ice, but the urinal is probably the safest, since if you threw it out the back door, some unsuspecting animal or stupid human might be endangered.

We talked to many restaurateurs and bartenders about this Imponderable, and found many that admitted to using urinals as dumping grounds for contaminated ice. Toilets, sinks, alleys, and the street were all favorite burial grounds for unwanted ice. Dumping ice mixed with broken glass anywhere in a public bathroom doesn't seem smart; most bars and restaurants try to melt the ice with hot water in a sink or service bin and then throw the glass in the garbage. But ice is more often rendered unusable by spilled drinks or small pieces of food—both can easily be flushed away.

But we suspected more was at work. Most urinals are not large enough to hold the volume of ice held in bins behind bars. We've also seen ice put in urinals in rest rooms in gas stations without air conditioning, especially in the desert. Said bathrooms have a tendency to, let us be kind, not be fragrance free. We side with Gunnar Baldwin, senior national accounts manager of urinal manufacturer TOTO, USA, who wrote us:

> The ice is put there to act as an automatic flushing system. As urine melts the ice, enough water is produced to help the urine siphon out of the trap within the urinal. Of course, the melting rate is proportional to the surface area of the ice, so crushed ice works better than larger cubes.
>
> It also has the benefit of preventing splashing and reducing odor. Cold urine does not emit as much odor! But the main purpose is to melt enough ice to carry the urine down the drain.
>
> The restaurant doing this is likely to have had problems with men who do not want to touch the handle on the flush valve. To use ice in such volumes is expensive. They would save quite a bit in a short time by installing our sensor-operated flush valves, which need no touching, flush effectively after every use, are conservative with water, and create the image they seek in keeping their rest rooms clean for their customers.

Sometimes you can't get a good answer without a commercial plug!

Some folks argue that ice provides a pleasing target for men. Men can't resist target practice. Some men try to melt as much ice as possible, or to create mini-ice sculptures. Holland's Schiphol Airport installed urinals with small "targets," life-sized sculptures of flies (these can be seen at http://www.urinal.net/schiphol/), that according to a *Wall Street Journal* article, reduced spillage as much as 80 percent.

But even when men manage to hit the target, urinals will smell unless there is a steady stream of water to remove urine. Automatic-flush urinals are probably the best answer, but many establishments are saddled with old plumbing. The salt, sulfur, and other minerals contained in urine both harden in the trap and constrict the available "throat" of the trap of the urinal. This plumbing equivalent of "hardening of the arteries" can cause problems even on those rare occasions when men actually do flush voluntarily, creating dreaded backsplash.

But the slight inconvenience to customers or the long-term plumbing problems are unlikely to be what motivates the dumping of ice in urinals. As one bartender put it: "It's the smell, Dave. It's the smell."

Submitted by Judith and Paul Dahlman of New York, New York.

Why Do Auto Batteries Lose Their Charge When Left on a Concrete Floor?

It's probably not a coincidence that the two readers who posed this Imponderable come from cold-weather climates. It turns out that car batteries seem to like the same temperatures we do when we're in our shirtsleeves, as Celwyn Hopkins, executive secretary of the Independent Battery Manufacturers Association, explains:

> A lead-acid storage battery loses its capacity at any temperature lower than 80 degrees Fahrenheit, since the electrochemical action is slowed down. At 32 degrees Fahrenheit, the battery is only 65 percent efficient, and at zero degrees Fahrenheit, it is only 40 percent efficient. When warmed back to 80 degrees, it would be 100 percent efficient, or fully charged.

> Hydrometers used for checking the specific gravity of battery electrolyte (state of charge) have a temperature compensation chart to arrive at the correct figure. If you took a fully charged battery at 80 degrees and gave it a hydrometer test, it would read 1.265, but at 30 degrees, it would read 1.245.

Hopkins adds, ominously: "There have been a lot of arguments on this subject and I usually don't win." Not exactly what we want to hear from a knowledgeable source!

A concrete floor in a garage is no colder than a wooden one, but it feels colder, mostly because concrete's greater density conducts cold (and heat) more efficiently. The naysayers concede that batteries don't thrive in cold weather, but argue that the differences in conductivity between concrete and wood floors or shelving are not sufficient to explain what they consider to be a long-held myth about batteries. Most agree with Gale Kimbrough, the technical services manager of Interstate Batteries, that although the premise of this Imponderable is no longer true, it does have a historical basis:

> Many, many years ago, wooden battery cases encased a glass jar with the battery inside. Any moisture on the floor could cause the wood to swell and possibly fracture the glass, causing it to leak. Later came the introduction of the hard rubber cases, which were somewhat porous and had a high carbon content. An electrical current could be conducted through the container if the moist concrete floor permitted the current to find an electrical ground. The wise advice of the old days to "keep batteries off concrete" has been passed down to us today, but it no longer applies because of the advanced technology of today's batteries.

Modern auto batteries are now encased by polypropylene, which insulates the battery much more effectively than rubber. Even so, Kimbrough points out that even at the ideal 80 degrees, "some lead acid batteries discharge 4 to 8 percent per month, and can lose even more in very hot weather. In fact, some garages have been known to put batteries in hot climates on concrete floors during the summer, just to keep the batteries cooler and retain their charge better."

Submitted by Frank Buller of Brewster, New York. Thanks also to Michael Javernick of Colorado Springs, Colorado.

Why Does Gum Get Hard When You Drink Water While Chewing?

Most food softens when moistened, but chewing gums stiffens. What's the deal?

Calling chewing gum a "food" is a stretch. Until about sixty years ago, most gum base was made from the sap of a Central American sapodilla tree—that sap was chicle (of Chiclets fame). In essence, folks were chomping on rubber.

By the 1950s, most major gum manufacturers replaced chicle with an artificial gum base made from a synthetic plastic-rubbery substance that chemically resembles the chicle it replaced. Although there are other ingredients in gum (sugar or artificial sweeteners, natural flavorings, glycerin to preserve moistness, etc.), it is the gum base that gives gum its characteristic elasticity and softness.

Chicle and the artificial gum base that was designed to mimic chicle share one important characteristic—they soften and harden over a small range of temperature. When moist gum cools, it hardens. When moist gum is warmed, it softens. When you stick a thermometer in your mouth, it registers 98.6 degrees Fahrenheit on your thermometer (give or take a degree or two or a hospital visit or two). When you chew gum, your near-100-degree saliva moistens the gum, and that rigid stick quickly softens.

When we drink water, it's usually cold. But even room-temperature water is cold enough to give our gum rigor mortis. Drink some hot water and the gum will magically soften up again.

This is not a chemical reaction. The gum doesn't care whether the cold liquid is Coca-Cola, water, ice, or malt liquor. And this is true of all rubber—pour some cold water on a rubber eraser and it will harden, too. In fact, as fans of Heloise know well, applying ice is the classic home remedy for removing dried gum from clothing—once you've hardened the gum, it is much easier to remove.

For the sake of research, we tested various water temperatures on a stick of Wrigley's Doublemint gum, and we found we could achieve stasis. When we opened the hot water faucet and the water was warm but not hot, we could drink it without the gum softening or hardening—the Golden Mean.

Submitted by Jill Clay of Pleasant Prairie, Wisconsin. Thanks also to Matt Weatherford of Arvada, Colorado.

Why Is There a Dot on Billiards and Pool Cue Balls?

You mean you didn't know that the dot was to cover up the nerve canal of an elephant? Doesn't anyone receive a proper liberal arts education anymore?

The earliest billiard tables and balls, created during the Renaissance in England, were made out of wood. Sometime during the seventeenth century, ivory balls were introduced. The British were already importing tons of ivory from Africa every year, and billiards players found the new ivory balls much more pleasing in weight, appearance, and sound (the lovely clicking noise when two balls collided).

But there was a serious problem with making a ball out of an elephant's tusk. Elephants have nerve canals running through the middle of their teeth, just like we do. To achieve an even, "honest" roll, the craftsmen who carved the balls so that the nerve canal ran straight through the center, creating dark imperfections at opposite ends of each ball. The ball crafters would usually plug the ball with something to assure even weighting. In the early days of billiards, ebony was often used to plug the canals, so that the holes appeared to be black dots.

By the early seventeenth century, ivory balls were more popular than wooden ones in England, and became the only balls used in serious competitions. But there were problems associated with these ebony-stuffed ivory balls. In his article in *Amateur Billiard Player* magazine, Peter Ainsworth explains:

> Holes created by the nerve would usually be plugged with ebony and become the "spot." Due to the general inconsistency of the spot ball and the tendency for it to "kick" when the ebony contacted the ivory of the object ball, it was considered to be a disadvantage to play with it.
>
> In addition to these problems, the porous ivory could also change shape during the course of a game as it absorbed moisture from a humid atmosphere. It was therefore common to see players when shooting from the balk [the line behind which you place the cue ball after an opponent "scratches"], carefully placing their ball so that the "poles" of the central nerve were exactly horizontal.

As they gained more experience in fashioning ivory balls, craftsmen realized that the ivory taken from near the base of the tusk was difficult to work with, as the nerve hole was wider than those nearer the tip. To assure equal weighting, some ball makers would demand only the center of the tusk in order to line up the nerve canal through the ball's center. In an e-mail to *Imponderables,* Peter Clare, whose family owns Thurston, one of the oldest and most respected manufacturers of billiard equipment, noted that top-quality ivory balls of this era had "very small evidence" of the nerve, sometimes insignificant enough to be covered by black dye rather than a solid material.

But the price of this expertise came high, and not just in terms of money. One tusk could yield material for only two or three balls. Elephants were being slaughtered to provide four or six billiard balls! According to Titan Sports, an English billiard-supply company, in the peak years of production, 12,000 elephants were slaughtered annually just to supply Britain with billiard balls.

In the late nineteenth century, plastic balls rolled to the rescue to supplant ivory. The first plastic balls were made out of celluloid, and later plastic resins, which, except for inferior acetate balls, are what most pool and billiard balls are composed of today.

So if modern pool balls are plastic, with no nerve holes in sight, why are there dots on balls today? Actually, not all cue balls do sport dots—not even the majority. But dots still appear on many balls, for a very practical reason. John Lewis, director of leagues and programs at the Billiards Congress of America, the governing body of pocket billiards ("pool") in the United States, explains:

> Most cue balls in pocket billiards do not have a dot on them. Some cue balls in pool are manufactured with dots, circles, or logos on them, but this is expressly so players can most easily determine which make of cue ball it is. . . . When dots, circles, or logos are stamped on cue balls, it is because the white surface is ideal for marking a ball with a manufacturer's identification mark. It has nothing to do with the evolution of the cue ball with the natural dot from ivory times.

In other words, the all-white cue ball is the best possible "billboard" for an advertisement for the manufacturer, just as the white space on the ace of spades provides the requisite white space for a plug for the card maker.

The most popular carom billiard games (featuring tables without pockets), such as English billiards and three-cushion billiards, are played with only three balls: one red, and one cue ball for each of the two players. A player is not allowed to shoot the opponent's cue ball, so it is important that each player be able to easily identify whose cue ball is whose. Usually, one cue ball is pure white; the other has a dot, a colored circle, or a logo to distinguish it from the other. More often than not, there are two dots on billiard cue balls, so that the mark is distinguishable if one dot is flush against the table.

Anyone who has seen *The Graduate* or listened to Frank Zappa can attest to the poor public relations that the plastics industry has endured. But the pool and billiards industry is mighty pleased with its adoption of plastic balls, which are cheaper and easier to manufacture. Players are happy with their perfect roundness and true roll. And elephants are downright ecstatic about them.

Submitted by Patricia Roberts, via the Internet.

Why Are Some Parts of Our Bodies More Ticklish Than Others?

In our first book, *Imponderables,* we explored why we can't tickle ourselves, and noted that the neural pathways that control tickling are identical to those that cause pain. So the experts who tackled this Imponderable focused on serious benefits that ticklishness might bestow on us mortals, all agreeing that what we now consider a benign tingling sensation at one time in our evolution might have warned us against serious trauma.

San Francisco biophysicist Joe Doyle notes that some parts of our body are more richly endowed with nerves than others—including such tickling meccas as the bottom of the feet, the underarms, and the hands and fingers. Evolutionists, notes Neil Harvey, of the International Academy for Child Brain Development, "would say that the reason for the heavier concentration may be whatever survival benefits we derive from being more sensitive in those places."

How could, say, the armpit possibly be necessary to survival? "The axilla warns of a touch that might progress to a wound of the brachial plexus, which could paralyze an arm," answers University of Chicago neurosurgeon Sean F. Mullan. Other sensitive sites such as the nostrils, ear canals, and eye sockets are all subject to invasion by foreign objects or creeping or flying insects.

What about the underside of the foot, then? Mullan is slightly more tentative:

> The role of the foot is more perplexing. Is it a warning against the snake that crawled up the tree when we lived in its branches? Is it a hypersensitivity resulting from the removal of the thicker skin of our soles, which was normal before we began to wear shoes? I prefer the former explanation.

Submitted by a caller on the Mike Rosen Show, KOA-AM, Denver, Colorado.

If All Time Zones Converge at the North and South Poles, How Do They Tell Time There?

Imagine that you are a zoologist stationed at the South Pole. You are studying the nighttime migration patterns of Emperor penguins, which involves long periods observing the creatures. But you realize that while you watch them waddle, you are in danger of missing a very special episode of *The Bachelor* on television unless you set that VCR for the right time. What's a scientist to do?

Well, maybe that scenario doesn't play out too often, but those vertical line markings on globes do reflect the reality. All the time zones do meet at the two poles, and many *Imponderables* readers wonder how the denizens of the South Pole (and the much fewer and usually shorter-term residents of the North Pole) handle the problem.

We assumed that the scientists arbitrarily settled on Greenwich Mean Time (the same time zone where London, England, is situated), as GMT is used as the worldwide standard for setting time. But we found out that the GMT is no more! It is now called UTC (or Coordinated Universal Time—and, yes, we know that the acronym's letter order is mixed up). The UTC is often used at the North Pole as the time standard, and sometimes at the South Pole.

We veered toward the humanities in school partly because the sciences are cut and dried. If there is always a correct answer, then teachers could always determine that we came up with the wrong answer. Science students were subjected to a rigor that we were not.

But when it comes to time zones, the scientists at the poles are downright loosey-goosey: They use whatever time zone they want! We spoke to Charles Early, an engineering information specialist at the Goddard Space Flight Center in Greenbelt, Maryland, who told us that most scientists pick the time zone that is most convenient for their collaborators. For example, most of the flights to Antarctica depart from New Zealand, so the most popular time at the South Pole is New Zealand time. The United States' Palmer Station, located on the Antarctic Peninsula, sets its time according to its most common debarkation site, Punta Renas, Chile, which happens to share a time zone with Eastern Standard Time in the United States. The Russian station, Volstok, is coordinated with Moscow time, presumably to ease time-conversion hassles for the comrades back in Mother Russia.

Early researched this subject to answer a question from a child who wondered what time Santa Claus left the North Pole in order to drop off all his presents around the world. Based on our lack of goodies lately, we think Santa has been oversleeping big time, and now we know that time-zone confusion is no excuse.

Submitted by Thomas J. Cronen of Naugatuck, Connecticut. Thanks also to Christina Lasley of parts unknown; Jack Fisch of Deven, Pennsylvania; Dave Bennett of Fredericton, Ontario; Paul Keriotis, via the Internet; Peter Darga of Sterling Heights, Illinois; Marvin Eisner of Harvard, Illinois; and Jeff Pontious of Coral Springs, Florida; and Dean Zona, via the Internet.

How Do You Tell Directions at the North and South Poles?

You think time zones are a problem, how about giving directions to a pal at the South Pole. By definition, every direction would start with "Head north."

In practical terms, though the distances aren't great at the science stations, and it's not like there are suburbs where you can get lost. But scientists do have a solution to this problem, as Nathan Tift, a meteorologist who worked at the Amundsen-Scott South Pole Station explains:

> If someone does talk about things being north or south here, they are most likely referring to what we call "grid directions," as in grid north and grid south. In the grid system, north is along the prime meridian, or 0 degrees longitude, pointing toward Greenwich, England, south would be 180 longitude, east is 90 degrees, and west is 270 degrees. It's actually quite simple. Meteorologists like myself always describe wind directions using the grid system. It wouldn't mean much to report that the wind at the South Pole always comes from the north!

Submitted by Michael Finger of Memphis, Tennessee.

Why Does Orange Juice Taste So Awful After You've Brushed Your Teeth?

Catherine Clay, of the State of Florida Department of Citrus, offers advice that is hard to refute:

> Most dentists suggest it is better to drink orange juice first, rinse the mouth with water, then brush the teeth, since we should brush *after* eating or drinking rather than before. Based on personal experience, I can tell you that drinking the orange juice prior to brushing seems to reduce the terrible taste problems considerably.

Flawless advice, Catherine, but where's your sense of danger?

For those of you who have never walked on the wild side, you've probably experienced a lesser version of the "toothpaste-OJ syndrome." Perhaps you've followed a heaping portion of sweet cake with lemonade and thought that someone forgot to put sugar in the lemonade.

The toothpaste-OJ syndrome works in reverse, too. We've always found that oranges taste particularly sweet after we've crunched on a pickle. These kinds of "flavor synergies" can work for bad or good (oenologists would argue that a Bordeaux and a medium-rare steak work together to enhance the taste of each), but aren't the same phenomenon as an actual chemical reaction. Toothpaste contains a chemical base, such as baking soda, while orange juice and other citrus fruits contain citric acid. The experts we spoke to were not sure of whether there might be a chemical reaction that would affect the taste of orange juice so drastically.

Clay cited the mint flavorings of most toothpastes as an offender as well:

> Eating a peppermint, spearmint, or other mint candy, then drinking orange juice results in the same problem. Also, most

toothpaste products are formulated to prolong the mint flavor to enhance the belief in long-lasting, fresh breath.

The most likely culprit in the particularly awful toothpaste-OJ synergy, though, is an ingredient in all of the biggest brands of toothpaste: sodium lauryl sulfate, or SLS. You'll find SLS not only in toothpaste, but in shampoo, shaving cream, soap, and, ahem, concrete cleaners, engine degreasers, and car wash detergents. What do all of these products have in common? The need for foam. SLS, a derivative of coconut oil, is a detergent foaming agent used to break down the surface tension of water and penetrate solids while generating prodigious gobs of foam.

Taste researcher Linda Bartoshuk, of the Yale University School of Medicine, notes that the active layer in the taste system is a phospholipid layer:

> You know what happens to a layer of lipid when you add a detergent to it? Well, that's what happens to your taste system when you put detergent in your mouth, brushing teeth. So you brush teeth and the phenomenon is that your ability to taste sweet declines, and everything that should normally taste sweet, tastes as if a bitter taste has been added to it.

SLS will also affect your perception of salty foods. If you eat salty snacks such as potato chips or pretzels after brushing your teeth, the salt taste will be faint or missing, but any bitter taste will be magnified.

If toothpaste-OJ syndrome is ruining your life, you can always search for a toothpaste in a health food store that doesn't contain SLS. Although we don't know of any country that bans SLS's use in toothpaste, a search on the Internet indicates that some folks are concerned about its harmful properties. Warnings abound that SLS can harm the skin, the eyes, the hair, and the immune system. We found expensive health-store brands that trumpet their lack of SLS, but even "homeopathic-style" Tom's of Maine toothpaste contains SLS.

Although SLS does help clean the teeth, there are many detergents that are as effective. But consumers believe that the more suds they can generate in anything from bar soap to shampoo to toothpaste, the more effective the cleaning will be. If that were true, we would all take daily bubble baths.

Submitted by Dianne Love of Seaside Park, New Jersey. Thanks also to Lisa Wahl of Hawthorne, California; Monica Sanz of McLean, Virginia; Lisa Granat of Kirkland, Washington; Ernie Capobianco of Dallas, Texas; Angela McCarthy of Martinsville, Indiana; Jon Grainger of Lexington, Massachusetts; and Tim Walsh of Sarasota, Florida.

Who Was Casper the Friendly Ghost Before He Died?

You can't blame someone for wanting to know more about the backstory of Casper. Restless ghosts are a dime a dozen. Poltergeists are scary. But you don't run into many friendly ghosts, and none so relentlessly affable as Casper.

We thought the billowy puff of friendliness originated in comic books, but we were wrong. Casper first appeared in a Paramount Pictures short cartoon in 1945, although at that point he didn't have a name. Casper might have been friendly, but his co-creators, Seymour V. Reit and Joe Oriolo, fought over who thought of the story of the "Friendly Ghost." Reit insisted he did, since Casper was based on an unpublished short story of his, and Oriolo was "only" the illustrator (Oriolo later went on to illustrate and produce 260 Felix the Cat cartoons for television).

By all accounts, the first cartoon didn't set the world on fire, but the second, "There's Good Boos Tonight," was released in 1948, and several more were created in subsequent years. Although Casper never gave Mickey Mouse or Bugs Bunny a run for their money, the chummy spook was Paramount's second favorite cartoon character after Popeye in the 1940s and 1950s. In these early cartoons, nothing whatsoever was said or implied about how Casper became a ghost at such a young age. As Mark Arnold, publisher of the *Harveyville Fun Times,* puts it: "They introduce Casper as a friendly ghost who doesn't want to scare people." Arnold adds that in the children's book that was a prototype for the cartoon, Casper's origins are undisclosed.

In 1949, Paramount sold the comic book rights to all of its cartoon characters, Popeye excepted, to St. John Publishing, which issued five Casper titles with a resounding lack of success. In 1952, Harvey Comics picked up the license. Harvey became Casper's comic book home for more than three decades. It was at Harvey where Casper was given a cast of sidekicks—his trusty ghost horse, Nightmare, and his antagonist, Spooky, the "Tough Little Ghost." Casper also became pals with Wendy, the "Good Little Witch," who spun off her own titles. The success of the Harvey comic books goosed the interest in made-for-television cartoons—more than 100 episodes were syndicated.

But despite the need for storylines for all these outlets, Casper's origins remained shrouded in mystery, and as it turns out, this was no accident. Sid Jacobson, who has been associated with Casper for more than fifty years, told *Imponderables* that when the company bought the rights to the Paramount characters, Harvey was more interested in the then more popular Little Audrey (a not-too-subtle "homage" to Little Lulu). Casper was thrown in as part of the deal, and he and other editors at Harvey went to work "rethinking him." Why the need to rethink? It turns out that Jacobson was less than thrilled with the original animated cartoon: "It was so ugly, and so stupid, I never forgot it. If we used the original premise for our books, it would have been a failure."

Ever mindful that Casper was meant to appeal to a younger segment of the audience, the editors at Harvey wanted to banish elements that would frighten children or give parents an excuse to ban their kids from reading about even a friendly apparition. Jacobson says:

> Since the dawn of the Harvey Casper character, truly the Casper everyone knows and loves, Casper's origin is definite but flies in the face of conventional definition: he was *born* a ghost. Like elves and fairies, he was born the way he was. We consciously made the decision as to his creation. It stopped the grotesqueries, and fits in better with the fairyland situation. It allows Casper to take his place with the other characters in the Enchanted Forest. It doesn't deal in any sense with a kid wanting to die and become a ghost. That was our main concern.

Considering the treacly nature of the comic book, inevitably a few impure types have speculated about the secret origins of Casper. Mark Arnold reveals a particularly startling one:

> The most notorious origin story appeared in Marvel Comics' *Crazy Magazine* #8, in December 1974, in a story called, "Kasper, the Dead Baby." In it, they show that small boy Kasper was killed by his alcoholic, abusive father. It's pretty gruesome, but bizarrely funny in a kind of strange way. Marvel has disowned the story, as they have tried to acquire the Harvey license.

In 1991, during *The Simpsons'* second season, the episode "Three Men and a Comic Book" speculates that Casper was actually Richie Rich (another bland comic book star of Harvey's stable) before he died. As Arnold puts it, "Richie's realization of the emptiness that vast wealth brings caused his demise."

Most recently, in the feature film *Casper,* there are allusions to the ghost's past (his father dabbled in scientific spiritualism), but no real explanation for what makes Casper so damned friendly and why he was snuffed out before his prime. Maybe the best theory comes from comic book writer and author of *Toonpedia* (http://www.toonpedia.com), Don Markstein:

> Personally, I always thought it was his friendly, open nature that did him in. His family apparently didn't do a very good job of teaching him about "stranger danger."

Submitted by Steve, a caller on the Glenn Mitchell Show, *KERA-AM, Dallas, Texas. Thanks also to Fred Beeman of Las Vegas, Nevada.*

LETTERS

Thank you for all the wonderful letters and e-mails you've sent since the publication of How Do Astronauts Scratch an Itch? *praising our efforts. But we won't be publishing those. This is the space for folks who have a bone to pick with us, sometimes to the point of wanting to take a pickax to our head.*

We don't have space to publish all the worthy additions and corrections to our labors, but even the most picayune criticisms are welcomed, and will lead to modifications in future reprints. Without further ado, let's embrace the abuse!

Some things you can count on. The swallows will return to San Juan Capistrano. Every summer we will be bombarded with crummy sequels to movies we didn't care about in the first place. And the Red Sox will field a promising team that will wilt in the clutch.

In this, our tenth volume, we wax nostalgic for some of the Imponderables that elicit the most passionate letters of comment. And of course, if there is an Imponderables *book, there will be letters about why ranchers hang boots on fence posts, a mystery first "answered" in* Why Do Clocks Run Clockwise?

For those of you new to the fracas, here are some of the theories that have been advanced to explain this phenomenon: to shield the post from rotting during rain; to discourage coyotes and other predators; to keep foul-smelling boots out of the house; to display pride; to mark where repair work on a fence is required; to amuse themselves; to signal that someone is home; to point toward a rancher's home (in case of heavy snowfall); to keep horses from impaling themselves on posts; to point toward the nearest graveyard; to shield posts from adverse reactions to the sun; and to do something *with single shoes that lie on the road. But*

Imponderables *readers have indomitable spirits. There are always new theories, like this one from Nicki Woodard of Rapid City, South Dakota:*

I was born in Nebraska and raised in South Dakota, and to us this is just common knowledge. In the Old West days, when a cowboy got a new pair of boots he would put his old boots out on the fence posts. That way if another cowboy had a pair of boots that were in bad shape, he could take the ones the first cowboy left if they were in better shape than his own. I guess they were into recycling back then, too.

Pam Dellinger of Ashdown, Arkansas, wrote to confirm that a passage we quoted from a Tony Hillerman novel was true:

Perhaps part of the answer to this Imponderable could have been answered by heading southwest. It is a practice by many, where ranches are still of some size, to follow the Navajo tradition of using the boot at ranch accesses to signal whether or not family members are home.

As the drive from county-maintained roads to a front door can be considerable, the practice saves many miles of riding at a snail's pace. Why travel all that way if you don't have to? The boot is always a cheap resource (we are never without at least one pair beyond human wear), and it takes no thought to turn it up or down as you come or go.

We have no idea why Steven Serdinsky of West Covina, California, would ever think that the National Museum of American History could possibly be a more authoritative source than we are, but Steven wrote them about this Imponderable and received a response from Lonn Taylor, a historian in the social history division:

Several scholars in the field of folklore and anthropology have written about this phenomenon, and the only thing that they can agree on is that it originated in western Nebraska in the mid-1970s. In fact, an Associated Press story published in the Emporia, Kansas, *Gazette* of April 23, 1979, quotes Nebraska folklorist Roger Welsch as saying that a Nebraska farmer named Jim Lippincott originated the practice in 1974. It has now spread across the Great Plains from Texas to Alberta. [Our own research led us to the conclusion that Henry Swanson, before 1974, originated the boot ritual in the same general area of Nebraska.]

Two reasons are consistently advanced for the practice: that it protects the tops of fence posts from water, which would otherwise get into the cracks of the post, freeze, and eventually cause it to split; and that the human scent on the boots deters coyotes and wolves from crossing the fence lines into pastures. However, Tom Isern of North Dakota State University, who mentions the practice in an article entitled "The Folklore of Farming on the North American Plains," *North Dakota History* 41 (Fall 1989), tells me: "There aren't any practical reasons for putting boots on fence posts. They don't deter coyotes, they don't preserve the posts, they don't mark anything. They just offer a message to passersby, a message having to do with identity and hard work. They are folk monuments, just like an old threshing machine on a hilltop."

Boots might not scare away a coyote but just about anything seems to frighten baby pigeons—at least we sure don't see them very often. In our first book, Imponderables, *we wrote about why we never see baby pigeons, and readers have sent us sighting reports ever since. Here is one of our favorites, from Nat Segaloff of Los Angeles, California:*

My friend Pam and I drove up the California coast to Monterey last week and stayed at a couple of inns. Under the eaves of one of them, just down from our window, was an illustration to one of your original Imponderables. Seen close-up, a baby pigeon looks not so much like a bird but like a fuzzy, bile-covered turd. Which is consistent.

No wonder baby pigeons hide from us—they can't stand the criticism. Actually, Nat sent us a photo, and truth be told, "fuzzy, bile-covered turd" is an apt description.

Speaking of unpleasant birds, in When Did Wild Poodles Roam the Earth? *we discussed why sea gulls congregate in parking lots. The professional bird experts seemed to think that the feeding possibilities were plentiful in parking lots, and that expanses of asphalt resembled the sandbars where gulls often congregate in beach areas. But David Moeser of Cincinnati, Ohio, noticed Kassie Schwan's cartoon illustrating this Imponderable (it depicted several birds poring over a map, with one apologizing to the others, with one sad gull crying: "Gee, I could swear a wetland habitat was here! Honest!"). Moeser was inspired by a trip to a shopping center on the day before Christmas:*

My [first] reaction was exactly the same as the suggestion your funny artist makes via the bird in the picture. I figured there may have been a wetland there before the shopping center was built. But on second thought, I think not. A better explanation is that the pavement, heated up by the sun (especially if it's blacktop), is a warmer landing spot for birds (and their feet!) than frozen grassland, with or without snow on it.

And I'd like to debunk the food theory. Of the thousands of gulls (yes, "sea" gulls, possibly from the Great Lakes or other water holes up Canada way), in several large groups of hundreds each, only a dozen or two nibbled at some birdseed put down in one spot. The rest just watched the nearby humans who were watching them. A small pond of water nearby was completely ignored, although surely they would have seen it from the air. Meanwhile, a group of several hundred gulls at the other end of the parking lot whirled around in what seemed to be some kind of social ritual, seemingly pointlessly flying in circles for minutes before rising up to cruising altitude and winging their way south. Why did they choose to congregate at the parking lot instead of the oodles of open space on real ground in the surrounding countryside? I think they just considered it a good place to rest.

Maybe gulls just like cruising the mall? Teenagers have been doing so for decades. But we have to admit that we love the theory proposed by William Stickney of Cresco, Pennsylvania:

I've been fortunate enough to emulate the soaring seagulls by flying sailplanes. Sailplane pilots stay aloft by flying in lifts (rising air currents) just as gulls do. Two common types of lift are ridge lift and thermal lift. Ridge lift occurs when wind is deflected upward by rising terrain. Ridge lift is common along shorelines where the sea breeze blows against a bluff or large sand dune. Thermal lift occurs when the sun heats the ground unevenly.

Dark areas, such as asphalt parking lots, absorb heat faster than surrounding areas with vegetation or lighter colored soil. The solar heat is transferred into the air above the dark surface, making it hotter than the surrounding area. This relatively hot air is buoyant and will rise in a column referred to by pilots as a "thermal." Cooler air from the surrounding terrain then flows laterally into the hot area to replace the air that has risen away. Now this air is heated, and the cycle repeats.

If the dark area is very large, the lateral wind velocity induced by the thermal can overwhelm the prevailing winds and produce continuous flow. The parking lot effectively becomes a huge solar air pump. Thermals from East Coast parking lots can rise as fast as 1,000 feet per minute, and can often lift sailplanes to altitudes of 6,000 feet or more. Seagulls, of course, can stay aloft in much weaker lift.

In my opinion, seagulls are attracted to shopping center parking lots in part because of abundant thermal lift, which allows the gulls to soar while looking for food, just as they do in nature.

When we answered "How Do Fish Return to a Lake or Pond That Has Dried Up? in When Do Fish Sleep?, *we focused on the ways that fish or eggs could "hide" in the supposedly "dead" lake and how fish can swim back once water has returned. But one enterprising reader, Paul H. Roek of Madison, Tennessee, wants us not to duck and cover, but to cover the ducks:*

Waterfowl, especially the plentiful mallard duck, will dine in an active lake on wild rice and other foods that contain fish and frog eggs. Then, flying to nearby lakes and ponds (whether they are recently rejuvenated by water or not) to rest, nest, or sleep, they "recycle" those ingested eggs of fish and frogs, which hatch and are now located in a new home.

Apparently this mystery was studied in "landlocked" lakes or ponds, those not having a stream, creek, or river flowing in or out. This made sense to me, and I have watched the eating habits of ducks, and found small lakes and ponds in my old home state soon had species of fish that were once not there. I had always guessed that some fisherman had caught the fish in one lake and released them in another.

Speaking of ducking and covering, martial arts exponents are still furious about what many of the boxing coaches said in Why Do Clocks Run Clockwise? *about why pugilists make a loud sniffing noise when punching. Rodney Sims wrote to us via the Internet:*

In martial arts, you are taught not only to exhale when delivering a blow, but to vocalize along with this exhalation when you wish to deliver a particularly powerful blow. In Tae styles, this explosive exhalation is *ke-ai* (pronounced "kee-eye"). It serves to focus the chi, which is one's inner power or spirit located at one's center (just behind your belly button) and push it through the extremity delivering the blow.

Mystical aspect aside, it does work. I speculate that the philosophy of mind, body, and spirit that must be present in order to excel at any coordinated physical activity manifests in different ways through the necessary translation from teacher to student. Finding a way to connect or relate to a youngling in teaching is sometimes difficult, not to mention reaching more than one.

The martial arts teach you that to control yourself, mind, body, and spirit, is to reach for perfection, and value is placed upon such control: involuntary functions can be controlled, more force can be delivered, and things outside of normal understanding can be understood. Martial philosophy aside, it seems logical to assume that so many people are taught to do it, and consequently practice this exhale, that it does work to focus one's mind. From my experience, boards are easier to break when you *ke-ai*.

Does Michael Tyson ke-ai *when he bites an opponent? While we muse about that, Ryan Pentoney shot us an e-mail that focused on the physical effectiveness of the sniff:*

I have been trained in the martial art goshin ru, and through my four years, I have been taught to exhale on all offensive and defensive maneuvers (punches, kicks, blocks, and some stances and movements). I am not familiar enough to speak for boxing specifically, but I can offer an explanation for fighting in general.

Ira Becker's stab at the question [the owner of world-renowned Gleason's Gym pooh-poohed the effectiveness of exhaling while delivering a punch] seemed rather senseless. . . . In general, the body receives the necessary oxygen through the normal breathing patterns and bodily responses observed under such increased activity (wider nostrils, increased adrenaline flow, and an increased heart rate, more relaxed blood vessels, and the constriction of many of the capillary sphincters leading to the digestive and other "non-essential" organs and systems to maximize the amount of blood oxygen to the muscles and other useful organs and systems). The amount of carbon dioxide and oxygen expelled or inhaled, or an inhibition thereof due to the sharp exhale, is negligible. The time spent sharply exhaling converted into an inhale period would not be constructive, seeing as how these exhale periods occur at the time the fighter is throwing a punch (in the case of boxing)—it would not be advisable.

The next time you're feeling energetic, try throwing a punch or two in succession while inhaling, and then while exhaling. You will probably find that it is harder to inhale while punching (holding your breath isn't very good either, as you inhibit gas flow altogether).

I believe that there is a greater purpose to the exhaling than simple gas exchange or a psychological reason. I have been taught that exhaling upon striking, blocking, exploding into a stance, or dodging out of the way of an attack severely minimizes the risk of having the wind knocked out of you. When you exhale quickly, your abdominal muscles tighten up and also protect your diaphragm. The opposite is true when you inhale. The end result is disastrous when you are struck in the upper abdominal area when you inhale, and in a fight can spell the end.

While writing about why power lines hum in How Do Astronauts Scratch an Itch? *we made one humdinger of a mistake. Readers Bill Schmidt and Harrison Leon Church echoed the sentiments of Bob Potemski of Mission, Kansas, who wrote:*

> On page 166 you define 1 Hz as one cycle per second (correct, of course), with AC at 60 Hz. In the next sentence you say, "The generator coil spins at sixty revolutions per minute." Shouldn't that be per second? Or have I failed physics?

Nope, we failed physics. This is one case where we blinded ourselves *with science.*

If our eyes are damaged, let's see if we can get in trouble with our ears. One reader agreed with our answer to an Imponderable about the effect on batteries when cassette players and radios are turned up to full volume (yes, they do indeed drain the batteries faster). But he was worried that some folks might have been misled by the discussion in Why Do Clocks Run Clockwise? *Frankly, we have not lost any sleep over the prospect, but we yield the floor to A. Wayne Hinson of Greensboro, North Carolina:*

> You said that the power is increased only 30 percent when cassette players are turned up to full volume, implying that the cassette player somehow would use less power and make batteries last longer than a radio. This is not the case. With a cassette player, the bulk of the power is used not to drive the speakers, but rather to drive the electrical motor and tape mechanism. The power actually used to drive the speakers would still increase by about 200 percent at full volume, but because the motor and mechanism use so much more power than even the maximum amount used by the speakers, the increase of the *total* power used would be a fairly low percentage, as you mentioned.
>
> For example, if the amplifier and speakers draw one-third

watt of power at low volume and one watt at full volume, but the motor draws one and two-thirds watts by itself, then total power consumption would increase from two watts at low volume to two and two-thirds watts at full volume, a 33 percent increase. A radio giving an equivalent amount of sound energy would use the same amount of power for the amp and speakers, but would not use any to drive a motor. So even though the increase in power for the radio is a far greater percentage, the total power consumed is far less, yielding drastically longer battery life at any volume, even full volume, than the cassette would have even at minimum volume.

Now that we've got that straight, the cassette recorder has become a relic, an endangered species.

Actually, we would be an endangered species if Ann P. Mahoney of Charlotte, North Carolina, had her way. She was annoyed with our discussion of why silos are round in When Do Fish Sleep? *We quarreled with the notion that the shape had anything to do with forestalling spontaneous combustion, but Mahoney asserts otherwise:*

In *When Do Fish Sleep?* you pooh-pooh the idea of spontaneous combustion taking place in a silo. I can tell that you aren't a scientist. (Of course, you've never claimed to be one.) You are also obviously not a farmer who has lost a silo to spontaneous combustion. . . .

Riding stables often store grain and hay for the horses in a separate building away from the stable. That way, if spontaneous combustion does occur, they will only lose the grains and the building but not the horses. Most people don't know that grains can generate enough heat for spontaneous combustion to occur.

Ann, you correctly pegged us as never having lost a silo to spontaneous combustion. Actually, we didn't mean to suggest that grains never fell victim to spontaneous combustion, only that the shape of the silo didn't have anything to do with this problem. But, much as we hate to admit it, Mahoney does have a point. Grains most susceptible to spontaneous combustion are unevenly packed, with air pockets creating mold. Round silos are easier to pack tight. Reader Jason Backs of Houston, Texas, provides another reason why you don't see rectangular silos:

> A silo may reach more than one hundred feet high. The outward force exerted on the walls by the silage is tremendous. A silo built with flat walls would never be able to retain its shape. The walls would bulge out and most likely fail. Exceptionally strong, round silos are built by stacking interlocking blocks in a circular pattern. Steel cables are then tightened around the blocks. The force exerted by the silage is distributed evenly around the entire structure. Silos are actually less sound structurally when empty than when full.

O.K. We're not farmers, we're writers! But sometimes even that skill fails us. In How Do Astronauts Scratch an Itch? *we discussed why glasses sweat when filled with cold beverages. We included the role of heat and humidity in the process, and ended the chapter with the words, "This explains why glasses, as well as humans, tend to sweat more in the summer." In no way did we mean to imply that humans and glasses sweat for the same reason. In other words, the average farmer could have stated it more clearly than we did. As I. Getman of Woodmere, New York, wrote:*

> Cooler glasses sweat because they condense moisture from the humidity of the ambient air; people are perceived as sweating when they cannot evaporate the sweat into the ambient air fast enough, for whatever reason. These are opposite, not similar, phenomena.

We knew that. Really. Really!

Speaking of sweating, we mentioned whether swimmers sweat in What Are Hyenas Laughing At, Anyway? *We heard from Carole Hole of the Alachua County Library headquarters in Gainesville, Florida, who attended a sports seminar where this very Imponderable arose:*

> At one meeting, Dr. Robert Cade, a renal medicine specialist from the University of Florida College of Medicine, who was heavily into sports medicine, described experiments he did to decide that very question. He had his swimmers wear a device to retrieve moisture lost through respiration. He weighed them before and after swimming, subtracted the moisture lost via breathing and provided that they did lose unaccounted-for weight, which could be lost only via sweat.

Cade knows a little bit about sweat—he invented Gatorade (which, of course, was named after "his" football team, the Florida Gators).

They're not selling Gatorade in jugs yet, but in How Does Aspirin Find a Headache? *we asserted that the only reason for those little dimples in plastic milk cartons was to add structural integrity to the container. One former maintenance engineer in the plastics industry wants to add another reason. We yield the floor to Barry Peffer of Plain Grove, Pennsylvania:*

> The reason we had the discs in the sides of the bottles was for volume control. Those discs bolted into the aluminum mold that forms the bottle and could be changed or have a little bit machined off to adjust the volume of milk in the bottle.
>
> The USDA inspectors and milk sellers want the gallon of milk to come right up to the bottom of the threads in the cap. They have to fill it that full so consumers don't think they are getting cheated and not getting a full gallon. The [dairies] also naturally don't want to have to put a couple extra ounces in to make it full, then they are giving away product.

The variables in blow-molding the bottles—temperature of the plastic, temperature of the chilled water that runs through the molds that form the bottle, higher or lower humidity or temperature on any given day, a different truckload of raw material from a different lot number—all affect the shrinkage of the bottle over the first twenty-four hours of its "life."

So the dimples were there as an easy way to adjust volume. Our inspectors had to check the volume with graduated beakers every shift.

Reader Dennis Parrish agreed with our discussion of why ceiling fans get dusty in How Do Astronauts Scratch an Itch? *but wanted to add to the mix:*

Dust also collects on regular box fans, even though they spin much faster and are vertical. Here's the reason: As the fan blade turns, there is a thin layer of air that moves with the blades—essentially tagging along for the ride. More simply put, the air next to the fan blade is static. Dust diffuses into that static air and settles on the fan blade as if it were not even moving.

We weren't so sure about this theory, so we asked a physics type we know and asked him if he agreed with Parrish's comments. He replied:

Sort of. It's true that the very thin layer of air right next to the fan blades is fairly motionless. But I think this explains more why dust particles that are already clinging to the blades aren't thrown off more often, rather than explaining how the dust particles become attached to the blades in the first place.

To me, his depiction sounds like dust particles enter this "thin layer" and then just float around in there, occasionally clinging to the fan blade. But that layer is very thin. A dust particle would be within that layer only momentarily (less than a second), before it would either become attached to a fan blade or rejoin the turbulent boundary of air in front of the fan and be quickly blown away.

The situation can be compared to a bug hitting the windshield of a moving car. Most of the time, the bug will be blown by the airflow in front of the windshield away from the windshield. Occasionally, a bug will penetrate into the thin (relatively slow-moving) region of air just above the windshield and actually hit the glass. But the bug won't enter that thin region of air and then just hang around suspended there for a while. Similarly, a dust particle moving toward a fan blade will most of the time be blown away from the fan by the general airflow around the fan. Occasionally, a dust particle will penetrate into the thin layer near the blade and become attached to the blade (it clings to the blade mostly because of static electricity forces, I presume). But the dust particle won't just float around in front of the fan for a while. If it doesn't cling, the slightest movement away from the blade will blow the dust particle back into the turbulent airflow around the fan.

The relatively slow-moving air right near a fan blade, or a moving windshield, or a wing of a plane, for that matter, is due to "viscous drag," or as it is sometimes called, "skin friction."

When you're in our line of work, the last name you want to see in your e-mail inbox is "snopes." Snopes.com, also known as the Urban Legends Reference Pages, is our favorite place on the Web for debunking misinformation. So it pains us to admit that we think David P. Mikkelson, who with his wife, Barbara, are the team behind snopes.com, didn't buy the explanation for the origins of Baby Ruth that was supplied by the candy's maker (i.e., that the candy was named after President Grover Cleveland's daughter, Ruth), nor, as a reader suggested, that the candy was named after the granddaughter of the Williamsons, who created the formula for the candy. David Mikkelson writes:

The Williamsons did not "sell the recipe" for the Baby Ruth bar to the Curtiss Candy Company. The Baby Ruth bar came about when Otto Schnering, founder of the Curtiss Candy Company, made some alterations to his company's first candy offering, a confection known as "Kandy Kake."

Mr. George Williamson was the head of Curtiss's rival, the Williamson Candy Company, and thus the producer of the Baby Ruth bar's biggest sales competitor, the Oh Henry! bar. The notion that Williamson would be offering product names to his stiffest competition—or that Schnering would name his bar after a relative of his company's largest rival—is a bit far-fetched, don't you think?

If we thought clearly, we probably wouldn't have taken the candy maker's story as gospel. To read the Mikkelsons' full discussion of the Baby Ruth story, direct your browser to http://www.snopes.com/business/names/babyruth.asp.

But some of our readers with word origins on their minds have their heads buried in books, bless 'em. Danny J. Elek of Geneva-on-the-Lake, Ohio, read our description of the genesis of the term "Good Friday" and recollected this exchange in John Steinbeck's The Winter of Our Discontent, *which was set on Good Friday:*

Ethan Hawley: "Why do they call it Good Friday?"

Joey: "It's from Latin," said Joey. "Goodus, goodilius, goodum, meaning lousy.

We wrote a book about word origins, Who Put the Butter in Butterfly?, *but one of the words we discussed was a number, specifically "86," and why it means, among other things, that a kitchen is out of a food item or used as a code to eject unruly or nonpaying customers. Patrick S. Doyen of De Soto, Missouri, writes:*

You said that 86 was used in the lingo to denote that the

kitchen was missing a particular item. I found it an amazing coincidence that in the military's *Uniform Code of Military Justice*, Article 86 is unauthorized absence (AWOL). This could be a coincidence, but it wouldn't surprise me if the first cook to use this slang was an ex-Army mess cook.

For the curious, here's the Uniform Code:

> 886.Art.86. Absence Without Leave
> Any member of the armed forces, without authority
> (1) fails to go to his appointed place of duty at the time prescribed;
> (2) goes from that place; or
> (3) absents himself or remains absent from his unit, organization, or place of duty at which he is required at the time prescribed; shall be punished as a court-martial may direct.

Great catch, Patrick!

You know who could use better public relations? Public relations people! We received an e-mail from Craig Cherry of Eugene, Oregon, who was riled up about a number we quoted (in What Are Hyenas Laughing At, Anyway?) *from a representative of a dryer company, estimating how many trees are saved by the use of an automatic dryer in restrooms. Cherry was so distraught, we allowed him a horrendous "wooden" pun:*

> You quoted some statistics stating than an electric dryer in a public restroom would save the use of 125,000 paper towels a year, which weigh 906 pounds, or the equivalent of 7.7 trees.
>
> I don't know where this marketing director is from, but in Oregon, we don't cut our trees when they are 118-pound babies. (I'm assuming that paper is primarily wood fiber, and that wood can be made into paper efficiently).
>
> I think that paper towels are made of soft pulp wood, such

as cottonwood or poplar (among the pulp mill crowd, these are very poplar). According to Bruce Hoadley's excellent book, *Understanding Wood,* the specific gravity of cottonwood is about 0.35, or about 22 pounds per cubic foot; 906 pounds would be about 41 cubic feet of cottonwood, or 71,000 cubic inches. A ten-inch-diameter tree of this volume would be 904 inches tall, or 75 feet—a reasonable size for a tree. Spreading this volume among 7.7 trees of 10 inches in diameter means they would each be less than ten feet tall. The other trees would call them stumpy."

In How Do Astronauts Scratch an Itch? *we indicated that there were technical reasons why some cable companies rearrange the location of the VHF channels on their systems. But David Hanauer of Ann Arbor, Michigan, points out that the 1992 Federal Cable Act stipulated that those broadcast channels, especially noncommercial ones that meet the act's requirements, must be carried by local cable systems:*

> Those broadcast stations that request carriage and meet the Cable Act's qualifications must be carried on the cable system. The stations are required to be carried regardless of popularity. The stations have also been given the right, by law, to select their channel position on the cable system.

But you had other TV numbers in mind. In Why Do Dogs Have Wet Noses? *we mentioned that at one time there was a channel one. Dave Beauvais, an amateur radio enthusiast from Amherst, Massachusetts, heard us tackle this question on WBUR, Boston's NPR station, and wrote:*

> There was in fact a real-live TV Channel 1, which began broadcasting in Chicago in 1939. However, it was soon discovered that this frequency range (roughly 38 to 40 MHz) was so close to the international shortwave broadcast bands that it

was exhibiting shortwave propagation characteristics, rather than the predicted "line-of-sight" characteristics.

The manifest result was that Chicago's Channel 1 virtually blanketed the entire country from coast to coast. While it must have been thrilling for the few early experimental set owners in New York and Los Angeles to receive live television from Chicago, long before cable or microwave TV networks had been invented, this was clearly not what the FCC had in mind when it made the original channel assignments. The frequencies were meant to be "recyclable" in cities spaced roughly 150 miles apart. There was never any intention of giving Chicago a "nationwide" TV channel, which is precisely what happened!

When the allocation plan was revised after World War II, Channel 1 was in fact deleted and its frequencies given to police and fire utilities, which in 1938 had occupied the much lower frequency allocation between 1600 and 1800 KHz, just above our standard AM radio broadcast band. If you look at certain old household radios from the 1930s, you will indeed see that they extend to 1800 instead of 1600, and that the top end of the dial is labeled POLICE.

Speaking of police and numbers—get sent to the slammer and you'll be stamping out license plate numbers in prison. In What Are Hyenas Laughing At, Anyway? *we discussed how convicts got into the license plate business, and how states have been taking longer and longer to produce new plates. Instead, DMVs are issuing drivers renewal stickers, which are cheaper to produce. Reader Darrell B. Thompson of Los Osos, California, wrote:*

On page 139, you mentioned that "as yet, prisoners have not gotten into the renewal sticker business." Please note that the California Men's Colony (a state of California penal institution), located in San Luis Obispo, does in fact manufacture renewal stickers for the entire state of California. This operation, of necessity, is accomplished in a very secure environment in the prison and it is off limits to visitors and most of the prison staff.

We don't want to whine, but when readers get upset about letters of other readers, we get the mail. And in this case, they're up in arms about a number that doesn't exist—the 717 that hasn't graced a Boeing jet. Or has it? John Stancil, of Midwest City, Oklahoma, writes:

While reading *Do Penguins Have Knees?* I was surprised to find a letter [from a Boeing employee] explaining why there was no Boeing 717. I must point out there is more to the story.

I spent seventeen of my twenty-year Air Force career as a pilot of the KC-135, a Boeing-manufactured military air refueling and cargo jet aircraft. Every airplane I flew had a metal placard in the crew entry chute that bore its model designation as a Boeing 717.

And a devoted and passionate reader, Ken Giesbers of Seattle, Washington, just happens to be an employee of Boeing. He took umbrage with a reader in What Are Hyenas Laughing At, Anyway? *who claimed that the 707 number was assigned to Boeing by the FAA. We're on Ken's side, because he agrees with us!*

Your original answer was correct that Boeing chose the 707 designation. The registration number N70700 was requested, not forced upon Boeing. In fact, it was Carl Cleveland of Boeing Public Relations, the very department that your reader maligned, who convinced then-president Bill Allen to choose the number 707 instead of 700.

Ken was kind enough to contact Boeing's historical archivist, Tom Lubbesmeyer, who confirmed Ken's tale but extended an olive branch to the subject of Giesbers's wrath:

> Boeing public relations and sales people were planning to market the 707 under the name Jet Stratoliner. The selection of the number 707 (which was to be used primarily for internal record keeping) made obvious sense, because Boeing had historically made heavy use of the lucky number seven in its commercial model numbers since the Model 247 of 1934. And more specifically the name Jet Stratoliner harkened back to the original Stratoliner, the Model 307. Thus the 707 number quite nicely suggests the name Jet Stratoliner and vice versa. But once the 367-80 appeared, with the requested registration number N70700, it was the "seven-oh-seven" name that the public and press picked up on, and the Jet Stratoliner name was largely forgotten. So to a certain extent, the letter writer is correct in saying that the 707 was imposed on Boeing.

Lubbesmeyer then included smoking-gun evidence: a copy of the internal Boeing model register that shows the 707 number in September 1951, well before there was any need for any governmental registration number.

Speaking of registration, we heard from Doug Frazer of Cincinnati, who has "been known to drive eighteen-wheelers off and on since 1970." He confirmed what we wrote in When Do Fish Sleep? *that different states have all sorts of different regulations about where and whether state licenses must be displayed on the front and back of their trucks:*

> Some large tractors do indeed have rear license plates and rear license lights because some states require it. I have seen brand-new tractors from the factory with a license bracket light on the rear frame.

This rear plate is the same as the front plate, of course, and it's called the base plate. Since most tractors are engaged in the commercial transport of freight, the annual cost of a base plate can be thousands of dollars [in licensing fees to states].

In the case of a large corporation like Consolidated Freightways, the tractors may have a base plate from different states because the trucks "live" at different terminals nationwide. And most trucks have no rear plate—it gets so dirty under the trailer that such plates are solid mud anyway. In the case of a "shiny wheels" (owner operated), the base plate is usually from the state where the driver and truck reside.

It used to be that a truck running nationwide needed a blizzard of different plates, decals, signs, abbreviations, etc. This will never go away completely, because every state has its grubby paws out for money any way it can get it, and trucks seem to be easy targets. But in the last few years the states have been discovering that trucks travel in more than just their one state and some of the exterior junk is no longer required, because one decal now covers many states.

What most people don't see is the pile of paperwork in a folder in the cab: permits for all the states the truck will travel in. Modern owner-operators often use a laptop computer to keep up with all the paperwork, and there are plenty of software programs devoted to this.

And before we leave the subject of registration and transportation, we must certify one David Bennett, of Fredericton, New Brunswick, as being really real! The poor guy's letter was published in How Do Astronauts Scratch an Itch? *with one tiny little detail deleted:*

I noticed that you used my information regarding having to put snow into covered bridges. I'm glad you found it useful. However, I have one small request. You list me as "one enterprising reader from New Brunswick." Of course, nobody believes me when I tell them it's me! Is there any chance I could get my name mentioned when the book comes out in paperback? I'd get a whole bunch of free drinks from people who doubt my word!

Isn't it nice to know that your pals don't believe that you are enterprising? Or think you lack expertise about covered bridges? Please use this acknowledgment to cadge some free drinks from your wretched so-called friends!

Speaking of cadging, librarian Ethel C. Simpson of the Mullins Library at the University of Arkansas heard us on NPR talking about why pirates wear earrings. In How Do Astronauts Scratch an Itch? *we postulated that pirates believed that piercing their ears improved their vision. Simpson is suspicious:*

> I always heard that the earring was a way that a pirate could carry a little gold on his person—mad money or an emergency fund to tide him over if he got rolled or arrested.

We've heard this theory often, too, but we suspect that if a pirate were attacked, any gold on his person would be the first thing the roller would try to take from the rollee. Another listener to Weekend Edition, *Bruce Hope, e-mailed us with this theory that was advanced by a Navy tour guide on the USS* Enterprise *in San Diego:*

> I have heard that the earring was part of a sailor's uniform. If you have survived a shipwreck, then you have the right to wear a silver earring. If you have survived two shipwrecks, then you have the right to wear a gold earring. If you have survived three shipwrecks, then you are not allowed on another ship, as you are considered bad luck.

Three strikes and you're out, huh? We think that an enterprising sailor could foil this plan by the brilliant ploy of taking off the damn earring. Beats that problem of getting mugged for your jewelry, too.

But random muggings weren't the biggest threat to pirates—walking the plank wasn't exactly a walk in the park! In Astronauts, *we mused about why pirates forced prisoners to walk the plank instead of, say, just chucking them overboard. We mentioned several classic books about pirates in the nineteenth century that didn't mention walking the plank, including* Treasure Island. *One reader, Christopher L. Prenger of Bartlett, Illinois, wrote:*

> I am a recent discoverer of your fine books and have been poring over them voraciously and with much interest whenever I come across a new one. As always, I quickly read *How Do Astronauts Scratch an Itch?* and was again amused and informed. It so happens that at the same time I was in the midst of reading another classic of literature: *Treasure Island.*
>
> Your investigation into this matter, which I am sure was thorough as usual, I discovered was untrue on one small point. You state that popular pirate stories such as *Treasure Island* had no mention of walking the plank when, in fact, I found two references to it in the course of my reading. The passages are as follows:
>
> "1. His stories were what frightened people most of all. Dreadful stories they were; about hanging, and walking the plank, and storms of sea, and the Dry Tortugas, and wild deeds and places on the Spanish Main [chapter 1].
>
> 2. How many it had cost in the amassing, what blood and sorrow, what good ships scuttled on the deep, what brave men walking the plank blindfold, what shot of cannon, what shame and lies and cruelty, perhaps no man alive could tell [chapter 33]."
>
> Admittedly, the action of walking the plank never once occurs in the narrative but it is obvious that Robert Louis Stevenson was familiar with the concept, perhaps from the other classics of pirate literature you noted.

And while experts debate whether walking the plank actually existed outside of pirate tales, one naval historian and publisher of Time Rover Press, Dale R. Ridder, sides with the believers. He sent us a scan of a photo from a 1929 book, Our Navy and the West Indian Pirates, *that cites a shipping newspaper, the* Niles Weekly Register, *which documents a "walking the plank" incident in 1822, long past the height of piracy a century before.*

Let's switch to something slightly less life threatening—food and drink. We usually like to accompany liquids with some grub. But Mission, Kansas's Bob Potemski made us lose our appetite:

> I have a follow-up to the buffet plate question ("Why do many restaurants insist that you take a new plate every time you go through a buffet line?") in *Astronauts*. Even in restaurants where you need to use a clean plate for a buffet, you can still refill your drinking glass at the soda dispenser. Now, even if you buy the "germs from mouth to fork to plate to serving utensil to next person" scenario, it seems *much* more likely that since you put your mouth directly onto a glass, you could easily put germs onto the little metal arm that you press to get the soda. In turn, when the next guy or girl puts his or her glass on the arm, you get the picture?"

Unfortunately, yes.

While we're on the subject of gross dining practices, we might as well delve further into the subject of why pigs are roasted with apples in their mouths. In How Does Aspirin Find a Headache? *we admitted we couldn't track the origins of the practice. Jeannette Shelburne of West Hills, California, was researching Celtic mythology and found this passage in Barbara Walker's* The Women's Dictionary of Myths and Symbols:

> Today's Christmas pig with an apple in its mouth is the descendant of Norse Yule pig sacrifices, when the pig was of-

fered to the gods at the turn of the year. Pigs were holy in Germanic and Irish mythology, both branches of the long-established Indo-Aryan worship of Vishnu as the boar god. The Celts associated pigs with the other world and believed them to be the most appropriate food for sacred feasts. . . . Magic apples of immortality or of death and rebirth are common to most Indo-European mythologies. The apples are usually dispensed by the goddess to a man, hero, ancestor, or god. The Norse goddess Idun kept all the gods alive with her magic apples.

What goes well with apples, magic or mundane? Cheese, of course! In When Do Fish Sleep? *we noted that the only reason why American cheddar cheese is often orange is that consumers preferred the bright color. But reader Dan Chambers of Wisconsin Rapids, Wisconsin, wrote:*

Years ago, when cows were pastured, they sometimes had buttercups available as part of their diet. This gave the cheese a yellowish or orange color, and yielded a better flavor. Consumers noticed this and came to prefer the naturally colored cheese. It was a simple step to artificially color the cheese.

When we asked Dan for his source of information, he couldn't find the newspaper article where he first read it, but found a Web site based in the Isle of Mull in Scotland where a farmer relates:

One of our prime concerns . . . is to make in the winter and be able to sell in the summer. Winter, do not forget, lasts longer up here—from October through March—and the winter cheeses tend to be much paler in comparison to the buttercup yellow of the summer ones.

In What Are Hyenas Laughing At, Anyway? *we discussed why many guitar players keep long, dangling strings on their instruments. Patricia K. Larkins of Homewood, Illinois, seems to think we're dumb, which isn't exactly music to our ears:*

I am amazed that you could find eight reasons for the length of string on a guitar and still not come close to the right one. It's very simple: new strings stretch, and can slip off the post if they are cut too short too soon. That's all. Some musicians are particularly heavy-handed and are always breaking strings, so their strings never get old enough to cut off. I suppose that accounts for some copycats who think you should never cut the strings, but that's silly.

They should be cut as soon as they stop stretching because of the white noise they produce. Incidentally, a small piece of paper slipped under the strings just above the neck will reduce white noise and make your instrument sound much better.

All logical. But none of this explains the expanse of "extra" string on so many guitars, more than will ever be necessary.

Michele Myhaver's son plays the cello, and she has often pondered the mystery of why there aren't more left-handed string players. It's time to change the paradigm, according to contrarian Michele:

After watching my son, a lefty, learn the cello and many other people play myriad stringed instruments, I came to the conclusion that all stringed instruments are made for left-handed people. Take into consideration that a significant number of creative people are left-handed and that musicians are creative, and I think you will begin to see my logic.

The more complicated action in playing a stringed instrument is done with the left hand (i.e., compressing the strings and vibrating them for vibrato). All the right hand is doing is busily sawing away from the bow, a much simpler and repetitive motion. My son caught on more quickly to much more complicated fingering than any of his classmates because he had more control over his left hand.

In a world where the majority is right-handed and most lefties have a chip on their shoulder about being odd man out, the assumption is that all things were developed for righties. But I think this is a case in which the lefties may have had their way from the beginning, and by switching instruments around are actually making life more difficult for themselves instead of easier.

An ingenious theory, Michele, but there's one obvious flaw: if what you say is true, why aren't there a disproportionately large rather than small percentage of left-handed string players?

If lefties suffer discrimination, don't all humans compared to our "best friends"? Dogs have, well, a dog's life, as Jeff H. Johannsen of Sunnyvale, California, wrote in response to our foray into canine dental hygiene:

Why don't dogs get cavities? It hardly seems fair: We brush and floss two or three times a day and still we have to subject ourselves to sadistic Laurence Olivier [cool *Marathon Man* reference] impersonators while dogs eat anything that's lying about and never get a cavity. What's the deal?

We consulted our family veterinarian, Dr. Irwin Fletcher, and received an answer so simple and obvious we felt more than a little embarrassed.

Dogs, both male and female, have a gland that secretes a substance that is in many ways the chemical equivalent of fluoride. The gland is located directly below the anus. When your pet seems to be, er, polishing the family jewels, he is actually practicing good dental hygiene.

Our guess is that at best, the dog is multitasking, but who are we to argue about licking dogs when we have biting insects to discuss? In What Are Hyenas Laughing At, Anyway? *we quoted entomologists who believe that mosquitoes bite some people more than others because they prefer the smell of some humans to others. Wei-Hwa Huang, a longtime correspondent from Mountain View, California, thinks he has narrowed down their preferences:*

> Mosquitoes seem to be more attracted to distinctly human odor. Right after a shower, when you've washed all that dust and grime off yourself, mosquitoes are more likely to bite you because they can detect you more. Not to advocate avoiding baths, but in the past, when I've been with my sister, mosquitoes have tended to bite the one of us that took a shower last.

Just what we were looking for—another excuse to give folks for not *using soap and water. Please send all complaints to Wei-Hwa!*

Another olfactory issue inspired Elliot Ofsowitz of Sarasota, Florida, to write. He read our discussion of why snakes dart out their tongues in Why Do Clocks Run Clockwise? *We mentioned that snakes use their tongues as a tool for smelling and hearing as well as tasting. Ofsowitz focuses on the olfactory and, in particular, why snakes' tongues are forked:*

> Recent research and study on why most snakes have a forked tongue reveals that the snake's tongue picks up odor particles in the air and brings them into its mouth where, on the roof of its mouth the odor particles are sampled—sort of like smelling, only more accurate and precise. The fork in the tongue allows a differentiation between odors on the left and the right, sort of a 3-D smelling.

From pests to giants, another Californian, Jason Ly, wanted to add some information about why zoo elephants are often seen rocking from side to side. At the time he wrote the letter, Ly was a research assistant in the department of psychiatry at the University of California, Irvine, College of Medicine:

This type of rocking behavior is exhibited in both animals and humans and is known as stereotypy. Stereotypy is defined as the *excessive* production of one type of motor act, or mental state, which necessarily results in repetition that is deemed as "abnormal" behavior. This can be seen as the rocking back-and-forth motion often displayed by the mentally retarded, repetitive hand waving or flapping, and repetitive vocal sounds.

One probable cause of stereotypy is confinement. Some animals confined in small enclosures develop "caged stereotypies." These behaviors usually include rocking and locomotion, such as pacing up and down one part of the enclosure. This type of behavior can be broken by providing the animal with a big enclosure. People who reside most of their lives in institutions or within the confinement of prison exhibit a number of abnormal behaviors, many of which are repetitive. Although such behaviors are abnormal in form, this is a normal response to an abnormal environment.

Speaking of abnormal behavior, we pondered why Wayne Gretzky used to tuck in his hockey shirt. David Maxham shot us an e-mail:

The reason he tucked in his jersey, and only on one side, was so that when he was playing, the shirt wouldn't get in the way on the side that the stick was carried. The source of information comes from the Great One himself in his autobiography.

Great? Well, isn't it great that our Imponderable, "Why Are There No Purple Christmas Lights?" is moot? We received several purple Christmas light sightings. Based on this report from Robert Sherry of Mounds View, Minnesota, we have a feeling that our write-up in When Did Wild Poodles Roam the Earth? *is likely to become "un-moot" sooner rather than later:*

> There goes another theory shot to hell. There are purple Christmas lights, whole strings of them. More likely there *were*—I don't think they'll return next year.
>
> Target Stores carried strings of seventy pear lights labeled purple for this past Christmas season. Who knows if they were really purple or violet. I never saw them lit and expect very few others did, either.
>
> The after-Christmas sale, Target had a large quantity of purple lights available and even at 50 percent off, customers were hardly loading their carts with them. Purple lights are not without some redeeming social value—humor! I saw a couple looking at them, the husband trying to persuade the wife they were a good deal. She didn't say a word, and gave him the look—you know the one—"Am I going to have to put you in a home *already*?"
>
> The lights didn't appear this year. Who knows? Maybe we'll have pink lights next year?

Now as reluctant as we are to criticize our readers, it seems to us that some of their letters have a personal agenda. Matt Leveillee wrote to us all the way from Beijing, China:

> I would like a free copy of your book *How to Win at Just About Everything*, because I'm running out of snappy comebacks to stupid questions. I can't remember ever giving up on life—giving up is for losers. So I don't consider myself to be a loser.
>
> I entered a contest on Easy FM, my favorite radio station,

where you have to name your three favorite music superstars. If you do this, they promise you a box of Cadbury chocolates. Well, guess what? I never got any chocolates! So I wrote back to them and asked where my chocolates were. They wrote back, promising me free Wall's ice cream. I never got the ice cream, either! I'm mad, but I still listen to Easy FM because it's still my favorite radio station.

With this kind of loyalty to Easy FM, how can we not give Matt a free copy?

We do have a few readers who are loyal to us, though! In How Do Astronauts Scratch an Itch? *reader and musician Craig Kirkland helped us with a musical Imponderable. How did we repay his generosity? By misidentifying his instrument:*

I was highly pleased to find a paragraph of my writing and an acknowledgment in the back, but my little nit is that I play viola, not cello! A minor detail, but I like to spread the gospel for my neglected and underappreciated instrument whenever possible, so of course I was a tad disappointed.

I'd love to have another violist read the book, see my contribution, and feel the rising pride that a violist did something right instead of glorifying the grunting and guttural cello. Any chance that this error could be corrected?

There sure is. But why do we get the feeling we'll be receiving a few letters from cellists real soon?

We're willing to risk the wrath of cellists, but there's one group we don't want to tangle with—librarians! Laura Mae Leach of Moreno Valley, California, wrote:

I have just finished reading *How Do Astronauts Scratch an Itch?* and have enjoyed it very much, just as I have the previous books.

> I do have one quibble. In your preface, you state: "Our job is to track down the mysteries of everyday life that you can't answer by consulting reference books." As a public librarian, I must mention that even the question "Do Penguins Have Knees?" can be answered from several books at any self-respecting library. Of course, your answers are quite a bit more fun to read than most reference books, but I wish you would not mislead your readers this way. Now that many librarians are trained to navigate the Internet, libraries are an even greater source of answers to life's imponderable mysteries.

Fair enough. Just as Imponderables are ponderable ("Imponderability" refers to that state in which you think you will never find an answer), so are some Imponderables solvable if you know where to look in the library. This has always been the toughest part of deciding what Imponderables to include in our books—many mysteries are fascinating but solvable through traditional research in a library (or might be known to experts in a field but not published in mainstream sources). For example, in this volume, "Why Do Beavers Build Dams?" seems like an easy question to answer. Every documentary we've seen on beavers highlights their construction of dams. So in cases like this, we consult encyclopedias and other standard reference books to see whether the "whys" are answered. In this case, they weren't to our satisfaction. Because we were both interested in the answer to the question and clueless about its solution, we included this Imponderable. We're sure we could have found the answers in books, but not easily through standard reference works. We love librarians, and not just because they buy our books. They help us find our experts, and we feel a kinship for anyone whose pursuit is to track down the answers to annoying questions.

Leach concluded her letter with:

> You would make so many librarians and our customers happy if, as part of your tenth-book celebration, you would publish a union index to all ten books.

Who loves ya'? This has long been our most requested feature and we hope the back of this volume helps you unburden your library patrons of their Imponderable afflictions.

Well, at least one person loves us, and the feeling is reciprocated. Along with librarians, teachers are just about our favorite folks, especially this one—Larry Warmingham, now an administrator for the school district of Lancaster, Pennsylvania:

> I have spent the last ten years or so using your books in my classroom. I don't know whether this is a common thing or not, but I wanted to let you know that it has been very helpful.
>
> As a former middle-school science teacher, I was always looking for new ways to engage my students. I got your *Imponderables* book as a gift and was intrigued. I soon had many of your books on my shelf. It then occurred to me that if I wanted to get my students to be inquisitive and to look for answers that couldn't be found in the usual places, then I needed to ask them the right questions.
>
> The first question I used was, "Why is there a black spot in white bird droppings?" thinking that this would be a great start for hormonal preteens! It worked quite well. I used them once a week and gave prizes to students who were able to come up with the right answers. I worked it like the lottery. One set of prizes was divided among the correct answers. This way students were less likely to give away answers if they knew they would have to split the prizes.
>
> Students began to call Wendy's to find out why they had square hamburgers, e-mail M&M to find out what the letters stood for, etc. We even got to the point of creating a "final exam" at the end of the year. It's surprising how many students remembered the correct answers. The answers are not always in an encyclopedia, or a textbook, or a dictionary. Sometimes it's fun to look for the answer and very rewarding to finally find it. I think it was helpful to my students to search in new places,

to think outside the box, and to realize that answers can be found almost anywhere. Since I am no longer in the classroom as much, I have started to post the questions on my office door, and my colleagues stop by to give a guess to the answers each week. Just thought you'd like to know that there is more to Imponderables than just an interesting answer—but I guess you already knew.

Isn't it time to throw caution to the wind and develop an Imponderables *curriculum, Larry? We're available for consultation! Until you get back to us on that, we'll just have to bide our time, hunting down Imponderables before your students put us out of a job.*

Until we meet again, may you all jump for joy. Elephants would want you to.

ACKNOWLEDGMENTS

I wouldn't have been able to write a second *Imponderables* book without the support of readers, so it's hard to express the enormity of my gratitude for your contributions over ten volumes and almost twenty years. Your letters and e-mails do more than supply material for the books—they provide the energy to go on when I think of pursuing other alluring professions, like animal husbandry or envelope stuffing. I would say that I'm not worthy—but I don't want to give you ideas.

I'm so pleased to be back home at HarperCollins, especially with my editor, Susan Friedland, who loves to eat and laugh, two prerequisites for any cosmically attuned person. We first met sixteen years ago, when I used to loiter in her office; she finally realized it would be easier to edit me than to get rid of me. Thanks to Califia Suntree for her help.

Muchas gracias to two collaborators who have been part of my writing career since day one: my able agent, Jim Trupin, and the illustrious illustrator, Kassie Schwan.

For their invaluable research assistance, a big shout-out to Phil Feldman and Mark Sinclair, and a little whisper-out to Tom Rugg for his assistance on the Lone Ranger Imponderable. John Di Bartolo helped turn eleven indexes into one.

My friends and family deserve a medal for putting up with me, but they'll have to be content with a crummy acknowledgment. Mucho thanks to Fred, Phil, Gilda, and Michael Feldman; Michele Gallery; Larry Prussin; Jon Blees; Brian Rose; Ken Gordon; Elizabeth Frenchman; Merrill Perlman; Harvey Kleinman; Pat O'Conner; Stewart Kellerman; Michael Barson; Jeannie Behrend; Sherry Barson; Uday Ivatury; Laura Tolkow; Terry Johnson; Christal Henner;

Roy Welland; Judith Dahlman; Paul Dahlman; Bonnie Gellas; James Gleick; Cynthia Crossen; Chris McCann; Judy Goulding; Karen Stoddard; Eileen O'Neill; Joanna Parker; Maggie Wittenburg; Ernie Capobianco; Liz Trupin; Nat Segaloff; Mark Landau; Joan Urban; Diane Burrowes; Virginia Stanley; Sean Dugan; Alison Pennels; Marjan Mohsenin; Dennis, Heide, and Devin Whelan; Ji Lu; Alvin, Marilyn, Audrey, and Margot Cooperman; Carol Williams; Dan Fuller; Tom O'Brien; Susan Thomas; Tom and Leslie Rugg; Stinky; Matt Weatherford; and Amy Yarger.

Special thanks to my pals at Starbucks #839 for keeping me vertical; to John Di Bartolo, Annette Matejik, and my step-pals for keeping me ambulatory; to Jim Leff and Chowhounds for making sure I'm well fed; to PSML and Spectropop, for keeping the musical faith; to my Popular Culture Association pals, for getting academia right; to Bill and Saipin Chutima and Ali El Sayed for their artistry and friendship; and to the Housewife Writers for their frogspit.

And then there are all the experts in fields ranging from syringes to skunks who helped answer the Imponderables in this book. The most fun part of my job is when I find *the* expert or experts on some field I know nothing about, and hearing them talk about the subject they are passionate about. Without their willingness to share their knowledge with us, *Imponderables* would not be possible. My gratitude goes to all these sources whose expertise led directly to answers in the book:

Bob Allen, California State University, Fresno; Shirley Alvitre, Frank J. Zamboni and Company; Amurol Products Company; Robert Anderson, Idaho State University; Ron Anderson, Amana Division of Maytag Corporation; Lori Andrade, Stanley Bostitch; Cassie Arner, University of Illinois; Mark Arnold.

Connie Baboukis, Oxford University Press; Peggy Baker, Pilgrim Hall Museum; Gunnar Baldwin; Delia Barnard; Linda Bartoshuk, Yale University School of Medicine; William Benedict, Theatre Historical Society; Fraya Berg; Bruce Bjorkman, Traeger Grills; Deven Black; Michael Blakeslee, Music Educators National Conference;

Russ Born, Just Born; Stephen Brady; Julie Bridge, Thomas More Association; L. S. Brodsky; Fred Bronson, *Billboard*; Harold Brooks, Mesoscale Applications Group, National Severe Storms Laboratory; Shelor Brumbeloe; Stanley M. Burstein, California State University, Los Angeles; Peter Busher, Boston University; Brian Butler, U.S. Catholic Historical Society.

Campbell Soup Company; Lenore Campos; Bob Carrol; Bob Cate, Bowater, Inc.; John Chaneski; Amy Chezem, National Association of Chewing Gum Manufacturers; Faleem Choudhry, Pro Football Hall of Fame; John Churchill, Phi Beta Kappa Society; Peter Clare, Thurston Company; Warren Clark, Ford Gum Company; Catherine A. Clay, Florida Department of Citrus; Terry Collingham, Colonial Needle Company; Sean Collins, National Public Radio; Tom Conley, Lifoam Leisure Products; John Corbett, Clairol; Norman Cox, Franke, Gottsegen, Cox Architects; Sherri Creamer, Alive Again Bears; David Currier.

Doug Danks, Dome Corporation of North America; Lorraine D'Antonio, Religious Research Association; Thomas Deen, Transportation Research Board; Karen Del Principe, FASNY Museum of Firefighting; Alan Detscher, Secretariat for the Liturgy; John Di Bartolo, Polytechnic University; Michael DiBiasi, Becton-Dickinson and Company; Jim Dickinson, K-Tube; Pat Donahue, Kraft Foods, Inc.; Thomas Dorman, Dorman Publishing; Chuck Doswell, University of Oklahoma; Joe Doyle; Jerry Dragoo, Museum of Southwestern Biology, University of New Mexico.

Ray Ehrlich, American Plastics Council; EPS Molders Association; J. Richard Ethridge, Backyard Barbeques; Mark Evanier; Mark Evans, Purdue University.

Richard Fabry, *American Window Cleaner Magazine*; Fred Feldman; Phil Feldman; Arther Ferrill; Robert Fitts; Eddie Fizdale, Peak Produce; Scott Forman, New York Medical College, Westchester Medical Center.

Louis Galicia, Great Western Forum; Stan Garber, Selmer Company; Maura Gatensby; Burle Gengenbach, University of Minnesota;

Nat Gertler; Autumn Gill, University of Notre Dame; Jim Grady, Tri-State Window Cleaning; Janis Grant, North Alabama Wildlife Rehabilitators; David Gray, California State University; Robert D. Greenberg, FCC; Neil Grey, International Bridge, Tunnel and Turnpike Association; Mark Grunwald, Philadelphia Zoo; Gund, Inc.; Harmeet Guraya, USDA.

Jeffrey Hahn, University of Minnesota; Tom Hailand; Christopher Halleron; Richard Hanneman, Salt Institute; Ed Hansen, American Association of Zoo Keepers; M. N. Hartman, Vita Needle Company; Neil Harvey; R. C. Harvey; Gary Heinrichs, Wisconsin Department of Natural Resources; Frank Heppner, University of Rhode Island; Roger Herr, South's Bar; Billy Higgins, American Association of State Highway and Transportation Officials; Lauren Hiltner, Babe Farm; Keith Holmes; Celwyn Hopkins, Independent Battery Manufacturers Association; J. Benjamin Horvay; Michel Huet, International Hydrographic Bureau.

Sid Jacobson; Scott Jackson, University of Massachusetts, Amherst; Ben H. Jenkins; James Jester; Caleb Johnson; Carl Johnson; Lenworth Johnson, University of Missouri; Richard Allen Jones, Japan Information Center.

R. Kamins; Elizabeth Karmel, Girls at the Grill; Jane Kerber, Advanced Storage Technology; Arif Khan; Mindy Kinsey, *Teddy Bear and Friends*; Phil Klein; Dan Kniffen, National Cattlemen's Association; Mark Kohut, St. Martin's Press; Ken Koury; Bill Kretzschmar, University of Georgia; Jim Kunkel, C&W Co.

Delila Lacevic; Ella Lacey, Southern Illinois University School; Joseph Landau; Richard Landesman, University of Vermont; Susan L'Ecoyer, NARM; Dave Lewis, Corby Industries; John Lewis, Billiards Congress of America; Anthony Lojo, Swingline Staplers; Raoul Lopez, Culver City Ice Arena; Tom Lubbsmeyer, Boeing Co.; Lundberg Family Farms; Theodore Lustig, West Virginia University.

Soheila Maleki, USDA; Steven Marks, Wayne State University; Don Markstein, Toonpedia; Anthony Martin, National Flag Foundation; Garry Mauer, Window Cleaning Network; Mary Anne Mayeski,

Loyola Marymount University; Geoff Mayfield, *Billboard*; Linda McCall; Kevin McCray, National Ground Water Association; Jim McDade, University of Alabama at Birmingham; Melissa McGann, Charles D. Schulz Museum and Research Center; Rima McKinzey; Arnold Merkitch; Larry Meyers, MagTek, Inc.; Theresa Michael, General Crushed Stone Company; Don Monroe, Morton Salt; Robert Montgomerie, Queen's University; Stewart Montgomery, MagTek, Inc.; Michael Moore, Steinway & Sons; Dan Morrison; Arthur J. Mullkoff, American Concrete Institute; Sean F. Mullan, University of Chicago; Donna Myers, DHM Group.

Blake Newton, University of Kentucky; Masahiko Noro, Japan Foundation; Richard O'Brien; Bob O'Dell, Becton-Dickinson and Company.

Bill Palmer, Travel Centers of America; Jim Parham, Jive Records; Jim Patton, Smurfit Newsprint; Enid Pearsons; Doug Peters, Frank J. Zamboni and Company; Gin Petty; David Pickering; Janet Pope, Louisiana Tech University; Harrison Powley, Brigham Young University; Judy Provo, Purdue University; Rebecca Pyles, East Tennessee State University.

Quaker Foods and Beverages.

Real Foods; Recording Industry Association of America; Gary Regan; Rhino Records; Errol Rhodes, American Bible Society; Edward Richards, Louisiana State University; Charles Richman, EBM Industries, Inc.; Sara Risch, Science By Design; Andrew Ritter; Lloyd Rooney, Texas A&M University; Alan Rooscroft, San Diego Wild Animal Park; Steven N. Rosenberg; Jo Rothery, *Teddy Bear Times*; Tom Rugg; Russ Berrie and Company; Denis Ryan, Mine Safety Appliances Company.

Dan Scheeler, Sasakawa Peace Foundation; Tom Schott, Purdue University; Mike Schroeder, Amateur Hockey Association of the United States; Heidi Schwartz, *Today's Facility Manager*; Norman Scott, National Biological Service; Samuel Selden; Sherwood Medical Company; Carole Shulman, Professional Skaters' Guild of America; Brian Sietsema; Nina Simone; Skippy Peanut Butter; Dorothy

Skrotzki, Wildfur, Inc.; Whitney Smith, Flag Research Center; Louis Sorkin, American Museum of Natural History; Will Sousa, Babe Farm; Mike Starling, National Public Radio; Steiff North America, Inc.; Irwin Steinberg, Tortilla Industry Association; Dick Stilwill, National Appliance Parts Suppliers Association; Mike Stooke, Snookergames; Dennis Stout, E. D. Bullard Company.

Timothy N. Taft, University of North Carolina; Surnie Takagi; Alice Taylor, National Public Radio; Carol Thomas, University of Washington; Don Thompson, Comair Rotron; Nathan Tift; Carolyn Freeman Travers; Ritomo Tsunashima.

Father Kevin Vaillancourt, Society of Traditional Roman Catholics; Willy Voelzke, Rieger Printing Ink Company, Ltd.

Joe Walt, Sigma Alpha Epsilon; Ralph Waniska Texas A&M; Bobvin Ward, Society for the Preservation and Appreciation of Antique Motor Fire Apparatus in America; Janet Ward, Just Born; Robert L. Ward, Ahrens-Fox Fire Buffs Club; Lee Weaver; Weber-Stephen Products Co.; Marc Weinberg, Ballsy Bear; Brent Weingard, Expert Window Cleaner; Allison Wells, Cornell Laboratory of Ornithology; Lorraine Wettlaufer, Just Born; JoAnne C. Williams, Michigan Loon Preservation Society; Jon Williamson, North American Interfraternity Conference; Melanie Wong; Jim Wright, New York State Department of Transportation; Michael Wurtz, Sharlot Hall Museum.

Richard Zamboni, Frank J. Zamboni and Co.; and Mike Zulak, San Francisco Zoo.

And to the experts who preferred to remain anonymous but provided valuable research, a grateful tip of the hat.

INDEX

MASTER INDEX OF IMPONDERABILITY

Following is a complete index of all ten Imponderables® books and *Who Put the Butter in Butterfly?* The bold number before the colon indicates the book title (see the Title Key below); the numbers that follow the colon are the page numbers. Simple as that.

TITLE KEY

1=Why Don't Cats Like to Swim? (*formerly published as* Imponderables)
2=Why Do Clocks Run Clockwise?
3=When Do Fish Sleep?
4=Why Do Dogs Have Wet Noses?
5=Do Penguins Have Knees?
6=Are Lobsters Ambidextrous? (*formerly published as* When Did Wild Poodles Roam the Earth?)
7=How Does Aspirin Find a Headache?
8=What Are Hyenas Laughing At, Anyway?
9=How Do Astronauts Scratch an Itch?
10=Do Elephants Jump?
11=Who Put the Butter in Butterfly?

HELP!

If we haven't done it in nine other books, why should we fantasize that we have conquered Imponderability with this one? Ridding the world of annoying mysteries, raising the bar far higher than any elephant can jump.

So, please, continue to send in your Imponderables, your praise, and, yes, even your complaints. If you're the first person to send in an Imponderable we use in a book, you'll receive an acknowledgment and an autographed copy for your contribution.

Although we accept "snail mail," we strongly encourage you to e-mail us if possible. Because of the volume of mail, we can't always provide a personal response to every letter, but we'll try—a self-addressed stamped envelope doesn't hurt. We're much better with answering e-mail, although we fall behind sometimes when work beckons.

Come visit us online at the Imponderables Web site, where you can pose Imponderables, read our blog, and find out what's happening at Imponderables Central. Send your correspondence, along with your name, address, and (optional) phone number to:

feldman@imponderables.com
http://www.imponderables.com

or

Imponderables
P.O. Box 116
Planetarium Station
New York, NY 10024-0116

THE IMPONDERABLES® SERIES
by David Feldman
DO ELEPHANTS JUMP?
ISBN 0-06-053914-3 (paperback)
WHY DON'T CATS LIKE TO SWIM?
ISBN 0-06-075148-7 (paperback)
DO PENGUINS HAVE KNEES?
ISBN 0-06-074091-4 (paperback)
WHEN DO FISH SLEEP?
ISBN 0-06-092011-4 (paperback)
WHY DO CLOCKS RUN CLOCKWISE?
ISBN 0-06-074092-2 (paperback)
ARE LOBSTERS AMBIDEXTROUS?
ISBN 0-06-076295-0 (paperback)
HOW DOES ASPIRIN FIND A HEADACHE?
ISBN 0-06-074094-9 (paperback)
Available wherever books are sold, or call 1-800-331-3761 to order.
Don't miss the next book by your favorite author.
Sign up for AuthorTracker by visiting www.AuthorTracker.com.